Management for Professionals

More information about this series at http://www.springer.com/series/10101

Sanjay Mohapatra • Arjun Agrawal
Anurag Satpathy

Designing Knowledge Management-Enabled Business Strategies

A Top-Down Approach

Springer

Sanjay Mohapatra
Xavier Institute of Management
Bhubaneswar, India

Arjun Agrawal
Xavier Institute of Management
New Delhi, India

Anurag Satpathy
Xavier Institute of Management
Bhubaneswar, India

ISSN 2192-8096   ISSN 2192-810X   (electronic)
Management for Professionals
ISBN 978-3-319-81626-5   ISBN 978-3-319-33894-1   (eBook)
DOI 10.1007/978-3-319-33894-1

Printed on acid-free paper

This Springer imprint is published by Springer Nature
The registered company is Springer International Publishing AG Switzerland

Dedicated to
my family members
(Late) Dr. Baishnab, (Late) Dr. Suhila,
Hrishikesh, Kanyakumari,
Bharati, Sanjana, and Shrestha

—Sanjay Mohapatra

Dedicated to
my family members
Ajay Agrawal, Radha Agrawal
and Ambika

—Arjun Agrawal

Dedicated to
my gurudev Swami Shri Chidananda
Saraswati Maharaj, my family members
Basudev Satpathy, Urmila Satpathy,
Monalisa Satpathy, Sanjeev Padhy,
Raghunath Prasad Panda, Sujata Tripathy
and Dr. Manasi Panda

—Anurag Satpathy

Acknowledgements

The production of any book of this magnitude involves valued contributions from many persons. We would like to thank Neil Levine and Christine Crigler for providing continued editorial support and making this project a reality. Their association and patronage has become a motivational factor for us to write and publish with Springer.

We would also like to express our gratitude to many people who saw us through this book; to all those who provided support and assisted in the editing, proofreading and design—Gundeep Bindra, Himanshu Fauzdar, Gunveen Bedi, Varun Singhal, Vatsal Arya, Ankita Bhatia, Sudha Varanasi, Kaif Anjum Khan, Aakash Sethi, Bibhu Prasad, Shipra Khandelwal, Manoj Shukla, Vikram Singh Mahala, Atul Kumar Singh, Swapnajeet Padhi, Aditya Bikram Pattnaik, Bishnu Prasad Sahu, Risheek Raizada, Vikas Singh and Anshuk Pradhan.

Last and not least: We beg forgiveness of all those who have been with us over the course of writing this book and whose names we have failed to mention.

Sanjay Mohapatra
Arjun Agrawal
Anurag Satpathy

Contents

About the Authors

Sanjay Mohapatra received his B.E. from NIT Rourkela, M.B.A. from XIMB, M.Tech. from IIT Madras, India, and Ph.D. from Utkal University. Dr. Mohapatra has more than 29 years of combined industry and academic experience. He has consulted many organisations in different domains such as Utilities, Banking, Insurance and healthcare sectors. His teaching interests are in IT Strategy and Management Information Systems, and research interests are in the area of IT-enabled processes. He has authored/co-authored 21 books, more than 50 papers in national and international referred journals besides publications in different conferences. His contact details and list of publications can be found at http://ximb. academia.edu/sanjaymohapatra.

Arjun Agrawal completed his M.B.A. from Xavier Institute of Management, Bhubaneswar, India. He has varied experience in the field of Product Management, Banking and Information Technology. Prior to this, he completed his Bachelors in Engineering (Comp. Sci.) from Birla Institute of Technology, Mesra, India. He has served as reviewer for several conferences around the world and has also published several IEEE papers in India and abroad. His research interests include Marketing Strategy, International Marketing, Product Management, Knowledge Management, Information Systems and other related areas.

Anurag Satpathy is a consultant with one of the leading consulting companies in the industry. He has diverse experience in the field of handling consulting projects, Project Management, Quality Assurance and Technology Development. His research interests include Sustainable Development, Strategy & Operations, Innovation Management and Knowledge Management. Anurag holds a B.Tech. in Metallurgical and Materials Engineering from Malaviya National Institute of Technology, Jaipur, India, and an M.B.A. from Xavier Institute of Management, Bhubaneswar, India. His contact details and complete profile can be found at https:// in.linkedin.com/in/anuragsatpathy.

Chapter 1
Introduction to Knowledge Management

1.1 Learning Objectives

At the end of the chapter, the students would learn:

- Clear definition of knowledge management
- Key knowledge management concepts
- Different key concepts such as intellectual capital, communities of practice and organisational learning and memory

This chapter gives an introduction to the meaning of knowledge management (KM). It also provides a brief history on how knowledge management has become popular in the business world. A lot of KM actually existed before it started getting used as a strategic tool in business organisations. However, there is a lack of consensus as to what would constitute a good definition of knowledge management, but still the chapter provides a working meaning of knowledge management. The term KM is multidisciplinary and hence cannot be rooted in one particular function or department. Whatever it may be, there are two forms of KM—tacit and explicit. The chapter will deal with these two forms and how they change from one form to another and the importance of each form.

The usage of the term knowledge economy refers to an economy of knowledge which is used for production and management of knowledge. This usage is done in an organisation with constraints of the resources. The other meaning of knowledge economy refers to an economy which thrives on knowledge. In the second meaning, more frequently used, it refers to the use of knowledge technologies (such as knowledge engineering and knowledge management) to produce business benefits. The phrase was used by Peter Drucker in Chap. 12 of his book titled *The Age of Discontinuity*. The phrase was also used by an economist named Fritz Machlup. The essential difference is that in a knowledge economy, knowledge is a product, while in a knowledge-based economy, knowledge is a tool. Whatever may be the meaning, it involves multidisciplinary experts such as computer scientists, engineers,

© Springer International Publishing Switzerland 2016
S. Mohapatra et al., *Designing Knowledge Management-Enabled Business Strategies*, Management for Professionals, DOI 10.1007/978-3-319-33894-1_1

behavioural scientists, sociologists, etc. Some feel that knowledge economy is an extension of information era and hence feel that KM should be part of Information System department. The extension requires that business rules and practices, which are required for the success of any business model, should be written again such that the knowledge resources such as know-how and expertise are used as a base for solving any business-related challenges. According to analysts of the 'knowledge economy', these rules need to be rewritten at the levels of firms and industries in terms of knowledge management and at the level of public policy as knowledge policy or knowledge-related policy.

1.2 Difference Between Knowledge Economy and Traditional Economy

There are several discussions which try to find differences between knowledge and traditional economy. The consensus is that knowledge economy differs from traditional economy. The differences are with respect to several key aspects:

- In traditional economy, the resources are scarce and depleted. In knowledge economy, all the resources are not scarce but rather are in abundance. Unlike most resources that become depleted when used, information and knowledge can be shared and actually grow through application.
- The effect of location in knowledge economy is non-existent. In many of the activities in knowledge economy, by using appropriate technology and methods, virtual marketplaces and virtual organisations can be created. These virtual forums offer benefits of speed, agility and round the clock operation, and global reach can be created.
- Knowledge economy can create new business models by the creation of business clusters around centres of knowledge, such as universities and research centres. Even if clusters of knowledge already existed in pre-knowledge economy times, they never became a strategic tool for organisations.
- Knowledge economy helps to reduce barriers related to laws, taxes and geography. Knowledge and information are respected where demand is highest and the barriers are lowest. This economy does not follow traditional demand and supply phenomenon.
- Products and services which are rich in knowledge can command price premiums over comparable products and service that are low on embedded knowledge or knowledge intensity.
- In knowledge economy, pricing and value depend heavily on time of usage, content and context. Thus, the same information or knowledge can have vastly different values to different people or even to the same person at different times. In traditional economy, this variation is not seen.
- Intellectual capital, knowledge, has more intrinsic value than that of retail prices of products and services. Human capital competencies are key elements of value in a knowledge-based company. Of late, these capitals are reported in annual

reports in many 'learning' organisations, which give a lot of value to the knowledge contents in their employees.

- Information and communication architecture are necessary for knowledge flows. These are defined by social structures, cultural context and other factors which influence human behaviour. These are important to knowledge economies, which were not the case with traditional economy.

1.3   Understanding Knowledge Management

In knowledge economy, the ability to manage knowledge has become crucial to success. The ability to create, store and share knowledge has become ever more important factors to maintain the competitive advantage. More and more business houses regard knowledge as an important factor which is embedded in products (especially in products that are technology based) and as tacit knowledge, which is part of their existing employees. Knowledge is increasingly viewed as intellectual assets which have characteristics that are completely different than other valuable commodities.

1.4   What Is Knowledge

Knowledge is seen as a commodity that has the following characteristics:

- It is not reduced as it is consumed.
- It is not lessened when it is transferred.
- Knowledge can be created easily, but is difficult to be reused.
- When a knowledge transfer occurs, sum total of knowledge is more than it was before the transfer.
- Much of the knowledge is available in tacit form; as a result knowledge is lost when the employees leave an organisation.

With the onset of internet, the sources of knowledge have been unlimited and are abundantly available to all. After the industrial era, knowledge economy era has surpassed the industrial growth. In traditional economy, more than 50 % of the workers were involved in physical movements, who were known as traditional workers. According to a research, in knowledge economy, that percentage has come down to less than fifteen percent. The workers in traditional economy, who were largely in labour-intensive domain, have now given way to knowledge workers in relatively flat organisations. Earlier, the organisations had more layers making it hierarchical; but now the organisations are less hierarchical making them relatively flat. The knowledge workers have a different approach than the traditional workers. The knowledge workers require more collaboration, sharing of knowledge and know-how than their predecessors. The measure for success for the knowledge workers is how fast they can learn, acquire and share knowledge among themselves and then apply the knowledge in their organisations to sustain their growth. As per Davenport and Klahr (1998), a learning organisation is one that takes decisions based on knowledge that has been created, assimilated and diffused.

1.5   Need for Systematic Approach

All these developments in knowledge economy have mandated that managing knowledge cannot be based on individuals or groups alone. The knowledge management has to be systematic, based on scientific processes, so that the organisation can continue to be a learning organisation even after key resources leave the company. The need for systematic approach is to create, cultivate and share the company's knowledge base by storing valuable lessons and best practices learned from their day-to-day operations. The need also arises from the fact that the organisations need to learn from their past errors and not have to 'reinvent the wheel again'. However, organisation knowledge does not mean that it will replace individual knowledge but will add to it by making individual knowledge stronger, aligned towards achieving business goal, and making the knowledge more generalist so that many more in that organisation can use it. In the modern world, knowledge management is a deliberate attempt to ensure full utilisation of the organisation's knowledge base complemented with individual capability and potential so that an efficient, agile and productive organisation can be developed.

1.6   Definitions of Knowledge Management

Knowledge management (KM) is defined as a process-oriented approach to identify, capture, store, disseminate and then apply knowledge throughout the organisation so that the business transactions can be finished faster while being able to reduce cost of production as well as reduce rework. Another definition says that KM will provide solutions that will successfully help in making decisions for business purposes by using knowledge. These KM solutions will be explicit in nature and will be process oriented so that the performance can be repeated. These KM solutions are also known as intellectual capital management (ICM) solutions and are of business value to the organisation. These are also referred to as IC-based solutions. Many organisations also term them as intellectual assets. These assets are more visible in terms of patents, intellectual property, know-how, etc. ICM is different than the other KM solutions. In ICM, only the prioritised and best knowledge is stored and used.

There are several definitions of KM; these definitions have different perspectives which are given below. Table 1.1 gives a summary of different definitions as given in different available literatures. The various perspectives of KM are:

Table 1.1 Definitions from available literature

Davenport (1998)	KM is managing the organisation's knowledge base through a systematic and scientific process for acquiring, organising, sustaining, applying, sharing and renewing both the tacit and explicit knowledge of employees
Alavi and Leidner (2001)	KM is defining and implementing an organisation-wide process for enhancing organisation performance and to remain competitive
Allee (1997)	KM is a process of managing knowledge for effective decision-making process, thus creating value to all its stakeholders
Gupta et al. (2000)	KM is an organisation-wide approach that helps organisations to identify, accept, organise and transfer knowledge that will help improve day-to-day transactions
Holm (2001)	KM is finding ways to get the right information to the right people at the right time so that employees can create, share and reuse knowledge

- KM is about systems and technologies.
- KM is about people and learning organisations.
- KM is about processes, methods and techniques.
- KM is about managing knowledge assets.
- KM is a holistic initiative across the entire organisation.
- KM is not a discipline, as such, and should be an integral part of every knowledge worker's daily responsibilities.

From the definitions, it is understood that it is necessary to identify the knowledge that can be of value to the business and is also at the risk of being lost to the organisation. This loss can be because of employee turnover, retirement and competitors using ethical means to steal the knowledge. The definitions also imply that there will be three-tier knowledge that needs to be protected, viz. the knowledge at individual (creation of knowledge), group or community (sharing the knowledge) and organisation (reuse, adding value to stakeholders). The best way to retain valuable knowledge is to identify knowledge at the individual level and then to ensure that this knowledge can be retrieved and reused easily. This knowledge needs to flow from individual to individual and between members of a community of practice and then be diffused in the entire organisation so that the same can be made available to all and sundry. This flow can be in the forms of best practices, lessons learned, intellectual capital and organisation memory. Thus, there are four different phases in knowledge management—gathering, organising, generalising (making the knowledge more general in nature so that many more can use them) and reusing them. Figure 1.2 explains the four stages.

All these phases are designed with the following objectives in mind:

- To transition knowledge residing in the minds of employees who are going to retire to the employees who are going to succeed them
- To stop loss of organisation memory due to employee turnover and retirement
- To define process so that knowledge can be stored and saved from potential loss of knowledge
- To disseminate knowledge so that the same can be reused for efficient business transactions

1.7   Cross-Functions in KM

Knowledge management initiatives fail when they are considered as HR initiative or information system initiatives. KM involves several functions/departments and cannot be successful without integrated approach involving all stakeholders. KM has a foundation in each and every discipline/function/department of an organisation; using this foundation, these multidisciplines can be put to practice by quickly adapting to capture the knowledge of experts. After capturing these practices, they are reformulated as organisational practices, which eventually become organisation memory. This cross-functional approach of KM has been possible because it deals with knowledge as well as information. Knowledge is based on individual experience, expertise, perceptions and values. Thus, knowledge is derived from information by representing information in a more generalised way. This generalisation helps in better understanding, conceptualisation and further reuse. Information, in the first place, is derived by grouping data which are recorded from daily transactions. Thus, we can explain characteristics of knowledge management as follows:

Data: Numbers that are observed, measured, verified and recorded when a particular transaction happens are known as data. These are relatively raw and cannot be used for making decisions. Examples are sales numbers, number of products manufactured, etc.

Information: When raw data are grouped based on their characteristic(s), they represent analysed data. Examples are sales per month, sales per annum, sales per region, sales per demographic characteristics, etc. Similarly, units manufactured per month, units by category (such as number of detergent cakes manufactured per month, number of mobile phones manufactured per month etc.), represent information. Also, variance between plan and actual sales will be an example of information.

Knowledge: Information is benchmarked or compared with similar products in different business SBUs or verticals and organisations or with competitors' performance. This comparison is done to know the position of our organisation vis-a-vis other competitors. The result from this comparison is known as knowledge. When a decision is taken based on data and information, the chances of inaccuracies are high, whereas when the decisions are taken based on knowledge, the chances of accuracies are higher.

1.8   Types of Knowledge

In any organisation, there are two types of knowledge, viz. tacit and explicit. Tacit knowledge refers to the knowledge that resides in an individual's mind. This is usually difficult to articulate, draw or put in words. Explicit knowledge, on the other hand, will represent knowledge that has been recorded and represented in tangible forms. Tacit knowledge resides in heads of individuals; explicit knowledge is available in multimedia recording, paper format, etc. Table 1.2 explains the differences between the two types of knowledge (Table 1.3).

Table 1.2 Four major stages in knowledge management

Four major phases	Activities in each phase
Gathering	• Data entry
	• Pictorial/video input
	• Voice input
	• Pulling information from various sources
Organising	• Categorising
	• Numbering
	• Coding
	• Linking
Generalising	• Contextualising
	• Collaborating
	• Writing manuals
	• Describing procedure
	• Data mining
Reusing	• Seminars/best practice sessions/manuals
	• Sharing
	• Alert
	• Push through emails/templates/ procedures

Table 1.3 Explicit vs. Tacit Knowledge

Explicit	Tacit
Easy to diffuse, transmit, reproduce without ambiguity and share with others	Difficult to reproduce as the knowledge resides in an individual's mind and is context specific
Can be used for training and enhancing the skills as they are available in concrete forms	Abstract in nature and cannot be used for training because of lack of standardisation
The knowledge is categorised and coded and is easily available when there is a need	Cannot be made available all the time
Can be used for preparing user manuals and induction training for new joinees	Cannot be the base for training
The knowledge is standardised and hence can only be used for regular standard processes	Can only be used for new and exceptional cases
The knowledge transfer is good for groups and organisation as a whole	The knowledge transfer is good only for face-to-face discussion and knowledge transfer

Figure 1.1 shows similarities between tacit and explicit knowledge. We only get to see whatever we record in tangible forms such as multimedia, documents, etc. Implicit knowledge remains untapped in most of the organisations and will be lost if we do not define an effective KM system.

KM focuses on bringing tacit knowledge to explicit form and then defining methods by which the explicit knowledge will be applied in different parts of the organisation. Thus, KM will render tacit into tangible or explicit knowledge and then store the knowledge so that they can be retrieved and used when necessary. The key attributes of KM system are:

Fig. 1.1 Similarities
between tacit and explicit
knowledge

- Identifying knowledge
- Generating new knowledge
- Converting new knowledge into tangible form
- Sharing knowledge
- Embedding new knowledge in day-to-day business processes and procedures, user manuals and training kit
- Facilitating acceptance and growth of knowledge throughout the organisation by creating a learning culture and rewards and recognition scheme
- Measuring the benefits of knowledge in terms of business value and increasing their importance

1.9   Business Relevance of KM

An initiative such as KM can only be sustainable if it has business relevance. There are many business drivers, which have made KM initiatives successful and sustainable. The organisations where KM has been designed based on the business needs have succeeded in using KM as a strategic tool. Some of the business drivers are listed below:

- Organisations are operating across geographies, and the physical boundary lines no longer determine span of control. With globalisation being the key to success, integration of operations across multisite, multilingual and diverse culture operations is necessary for business success.
- The organisations need to work smarter and leaner. They do not need to 'reinvent the wheel'. They can reuse existing knowledge and build their business processes based on the knowledge gained. This will make the organisation lean and productive.
- Knowledge resides with individuals and thus is in their brains. When the employees leave the organisation because of retirement or otherwise, this knowledge

needs to be retained within the organisation. Fellow employees in their daily work can use the knowledge so retained, so that performance of the organisation can be improved. KM will address the issues of retaining this knowledge.

- Mistakes are avoided by learning from previous assignments shared with other employees. As a result, the past mistakes can be avoided resulting in lower rework cost.
- It will also help improve processes by creating working groups or community of practice. These group members not only acquire knowledge, but by using it in practical fields improve upon the defined processes by sharing their views and ideas. The improved process makes the organisation more competitive as they can face fast-changing dynamics of the business environment. Also the speed of solving the problems becomes faster.
- KM helps at the strategic level as both KM and business intelligence combined will help top management in strategic thinking and designing of competitive strategies. By using learnings and best practices in business (which are available in KM), the top management understands the present situation, risks and competitor threats. This also opens new doors for business opportunities as they are able to leverage on their present strengths while improving on the risks and threats.
- KM also increases customer satisfaction. The knowledge gathered through customer interactions can be fed back to a design team who can modify the design of products and services such that customers' implicit and explicit requirements are met. The manufacturing division then uses these knowledge-based designs to produce products which will result in increased customer satisfaction and subsequent loyalty. The customer support department also uses knowledge management systems to respond to customers' queries faster and solve their problems quickly. This knowledge would have been gathered, analysed and reused through well-designed KM systems which are aligned towards business operations.
- KM also helps in providing detailed information about different vendors and business partners which can be used for risk management. By using the past learning from different transactions, a vendor or a business partner can be considered for further business interactions and long-term relationships.

Today's business environment is quite complex, and there is a need for knowledge management to be an effective strategic tool to sustain the business model. There are thousands of information available leading to information overload. On a typical day, there could be fax messages, telephone calls, white papers, emails, etc. that contain the information required for running the business. Even though the need for KM is established, still, deriving knowledge from available information is not an easy task. Filtering emails and documents to derive context-specific knowledge is a business necessity. According to available literature, there are different approaches for KM which has been tried at different times. In the **first approach**, KM was considered as a tool-based initiative, and so more emphasis was given on information technology. This made alignment of business difficult with initiative, and most of the time, the initiatives did not get the desired business results. The initiative was more done in silos and integration with different departments often did not succeed. As a result 'best practices' and 'lessons learned' from different individuals were confined to a limited area.

In the **second approach**, KM focused on people and culture. The second approach started where the first approach ended. In the first approach, the reusage rate was low, and so the second approach gave emphasis on improving culture by making people more tuned towards learning from others and reusing the best practices. The approach was more of bottom up and groups were formed so that the learning culture can be propagated. These groups came to be known as 'communities of practice (CoPs)'. These groups helped in creating knowledge and then sharing it with other groups as well. These groups also took decisions to accept learning from other CoPs as well. But the problem in this approach is the lack of good mechanism for retrieval of required information. This era led to information overload. This fault led to the third approach.

The **third approach** took care of the pitfalls of the second approach, viz. information overload and retrieval of exact information. This approach increased awareness of the importance of categorising and coding contents. Based on the meaning of the contents, the categorisation will help in retrieving later. These metadata, which helped in describing the contents, were based on business necessities and had terminology from business requirements. In this approach, KM was considered to be collected in three phases—first, individuals will create knowledge; then, second, the CoP would validate it and help in categorising; and, finally, they will be shared with the entire organisation. These contents are usually coupled with metadata so as to give proper meaning and identification of the know-how.

1.10   Levels of KM

From the three approaches discussed in the previous section (business relevance of KM), we find that KM happens at three levels—individual, group (community of practice) and organisation. The knowledge is always created by an individual, which is reviewed and accepted in a community of practice. This CoP helps in validating the knowledge gained by an individual; the community practices it (maybe in pilot projects) and then, after a review process, accepts it. This is the period when the community helps to document the knowledge and record it in concrete forms such as documents, multimedia or prototypes. At this point of time, based on the third approach (as discussed in section 'Business Relevance of KM'), knowledge is categorised and coded and metadata are created. This helps in the easy identification, retrieval and indexing of the documents. After this stage, the stored knowledge is shared with the organisation using technology (such as emails, fax, bulletin boards or knowledge portal). Sharing the knowledge does not mean that KM has been successful. KM will be successful only when this shared knowledge is reused in the organisation for business transactions. Thus, we can make three levels of KM as follows (Fig. 1.2).

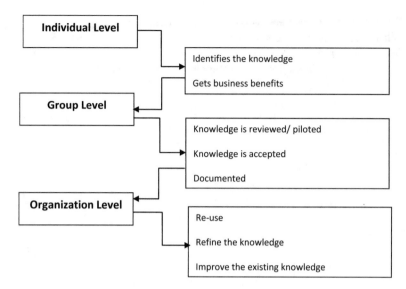

Fig. 1.2 Three levels of knowledge management

1.11 Summary

KM is not a new phenomenon or initiative, but has been in practice for a long time. The practice has been fine-tuned over a period of time so that KM initiative can be in tune with business goals. There are many definitions of KM, and this shows that the only way to succeed in KM is to make it part of all functions and discipline rather than keeping it to an elite few. Probably for this reason, KM is also considered as a multidisciplined initiative rather than belonging to a particular department. Knowledge is always generated in tacit form at individual level, and then it is practised and accepted at the group level called community of practice (CoP). Once the CoP accepts the knowledge, the same is converted to explicit form by recording in formal media such as multimedia, audio tapes, documents, etc. During this conversion, knowledge is categorised and coded and metadata are created. After conversion of tacit knowledge to explicit form, it is easy to share with the entire organisation and then reapply it elsewhere to get business benefits. Technology helps in sharing and reusing the knowledge but is not a constraint for designing knowledge.

1.12 Practice Questions

1. What is the difference between intellectual capital and knowledge?
2. How is knowledge created from data?

3. What is the difference between tacit and explicit knowledge?
4. What are the three approaches for knowledge management? How do they differ from each other?
5. What are the business drivers for knowledge management?

Chapter 2
Knowledge Management Cycles

2.1 Introduction

This chapter deals with knowledge management cycle. Any organisation which takes up knowledge management will undergo this process called KM cycle. The KM cycle shows us systematically how information is transformed into knowledge via creation and application process. When an organisation undergoes KM cycle then the following steps happen:

1. Capturing
2. Coding
3. Publishing
4. Sharing
5. Accessing
6. Application

There is no hard rule that only these steps should be there, but we are just trying to give an example. There are variations in different KM cycles so there might be other steps added or subtracted from the above list.

Businesses may follow any KM cycle says Wiig, a KM cycle or an integrated KM cycle or any other. Firstly, here we discuss the different stages in KM cycle and show how exactly this cycle takes place step by step. We also tell about the practical aspects of each step wherever possible. Secondly, KM cycle is explained in terms of transformation from tacit to explicit and then back to tacit. Thirdly, a comparison of pros and cons of different KM cycles is done, and finally, a comparison of where these can be applied is showed.

Traditional methods define business architecture as the relationship between people, process and technology. We agree with this definition as a starting point and add the dimension of strategy (Fig. 2.1).

The shortfall of this model is that it does not cover all the aspects of business so we will not be able to understand the linkages at the next level of detail.

© Springer International Publishing Switzerland 2016
S. Mohapatra et al., *Designing Knowledge Management-Enabled Business Strategies*, Management for Professionals, DOI 10.1007/978-3-319-33894-1_2

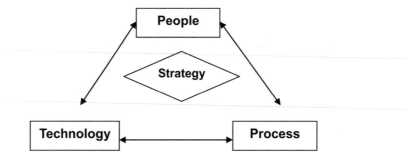

Fig. 2.1 Business architecture

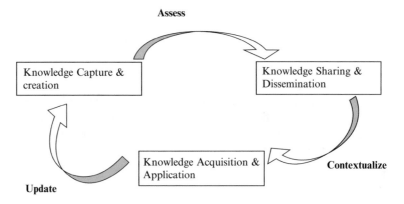

Fig. 2.2 An integrated knowledge management cycle

Process view tells about business process, the organisation chart and the management model.

Technology view models the inter-relational between the various applications, the data and the information flows and various service points.

People view tells about the inter-relational between the positions as defined in the organisation chart, the roles those positions fill and the specific individuals assigned to each position.

But how do we set up such a model?

We show how these stages take place by taking an example of a KM cycle. Figure 2.2 will show how some of the different stages take place in a KM cycle. First knowledge capturing is done and in transition from knowledge capture to knowledge sharing, knowledge content is assessed by other employees in the organisation or some other person outside the organisation. Assessment is done to know about the validity of the knowledge captured. After this the knowledge sharing is encouraged in the organisation. The knowledge is disseminated whenever the need arises in the organisation. After this the knowledge is contextualised. Contextualisation is the process in which the key attributes of the content are identified to match to the needs of the target users. For example, creating a short executive summary for the user so that he can quickly go through the document is contextualisation.

The knowledge is then used by the people in the organisation and then updated. During updating we check whether the knowledge is still useful or is it out of date now. Also, we look for the best practices and lessons learnt during update.

2.2 Different Stages in Knowledge Management Cycles

Capturing: Knowledge capturing refers to the identification and subsequent codification of existing internal knowledge and know-how within the organisation and external knowledge from the environment. This knowledge is usually previously unnoticed.

So what are things that a company should capture—the things that are of top priority and useful in future and not just everything? Focus on certain things only.

And the next question is who will capture those things? The answer lies in bringing in people in the organisation who have good writing skills, i.e. people like knowledge journalists, or who have done editorial work. These people have the knack of writing well and they are better than some Tom Dick or Harry.

How will he capture? It is by consolidating learning outputs and by taking feedback from peers. This is a practical way of knowledge capture in which we use a repository-based modelling tool for knowledge capture. This is nothing but some software where you can store the data in the form of say process diagrams, word document, etc.

Business process modelling tool features a multidimensional object repository that allows us to capture an object once and use it many times. Example of these objects would include things like process diagrams including their steps, the roles involved in executing those steps and the different technology types involved in supporting those process. The Web version of knowledge model was published in a modelling tool. This is what the employee would see when finding out something about their area.

Coding: The translation of valuable knowledge from tacit to explicit form is called codification of knowledge. Anything that allows the organisation knowledge to be communicated independently of its holder comes under the process of coding, for example, a document, a sound or video recording, a picture, a film, etc. These things are nothing but codification of knowledge. In most organisations the coded knowledge only represents just the tip of the iceberg because the tacit knowledge in huge is not coded.

During codification, we represent the knowledge in our minds by building mental models, i.e. by cognitive maps (which is defined below). We document the knowledge in manual and books in this step. Also the knowledge is encoded into knowledge bases.

When knowledge is codified it can know be communicated widely and with a less cost than what we had before codification.

Codification of knowledge can be achieved in many ways. Some are given below:

1. *Knowledge taxonomies:* It is a process by which the knowledge is partitioned and a relationship is defined between the partitioned knowledge. After this a classification is done. By this process we can understand the body of the knowledge better. An example of taxonomy is shown in Fig. 2.3.
2. *Cognitive maps:* A knowledge map or cognitive map is a representation of the mental model of someone's knowledge, and this provides a good way of codifying knowledge. A mental model is nothing but a symbolic representation of some knowledge in the real world. Cognitive map is a very useful way to codify the captured knowledge because it captures the complex relationships (see Fig. 2.4).
3. *Decision trees:* It is another good method to codify explicit knowledge. It is typically in the form of a flowchart. It has alternate parts which indicate the impact of different decisions. An example of this is shown in Fig. 2.5.

A few tools for doing knowledge creation and codification are:

1. For content creation:

   (a) Templates
   (b) Data mining
   (c) Blogs

2. For content management:

   (a) Metadata tagging
   (b) Archiving
   (c) Classification

Publishing: Knowledge is only useful if it is delivered to the right person, at the right time, in the right place. Knowledge publishing is defined as the process that allows getting knowledge to those people who need it in a form that they can use it. Different users may require the knowledge presented in a different way.

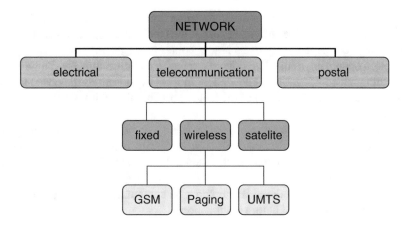

Fig. 2.3 An example of taxonomy (source: http://www.trainmor-knowmore.eu/)

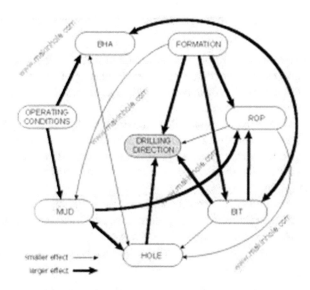

Fig. 2.4 Cognitive map to show the complexity of drilling (source: http://www.makinhole.com/)

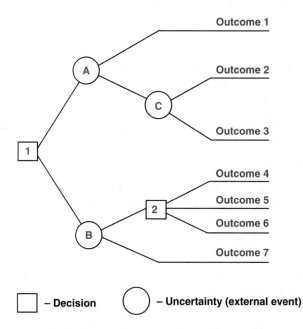

Fig. 2.5 An example of a decision tree (source: http://www.time-management-guide.com/)

Sharing: Once the knowledge is captured and codified in an organisation, it needs to be shared. The parties should be convinced that change in mentality is necessary for both the parties (the parties being the employee and the company). The employees should be told that sharing their knowledge would never affect their position rather it would earn a lot of respect and fame for him from his peers. Some tools for sharing are:

1. *Groupware and collaboration:* Groupware is nothing but a set of software that helps workgroups who are connected to each other by a network (e.g. LAN) to organise their activities and share knowledge. Groupware supports file distribution, electronic newsletters, etc.
2. *Wikis:* It is a Web-based software that allows open editing. This allows multiple users to edit and create content and is a good method of knowledge sharing, similar to Wikipedia perhaps in working. There is a word-like screen in which you can add your knowledge, and you need not even know programming or HTML commands for this.
3. *Networking technologies:* These consist of knowledge repositories, knowledge portals, Web-based shared workspaces, intranets, extranets, etc. Figure 2.2 which was shown above is an example of knowledge repositories in networking technologies. In this knowledge repository, knowledge is collected, summarised and integrated across sources.

Some methods by which employee would willingly share their knowledge are:

1. Make whatever they share more visible.
2. Make sharing a performance goal. This can be an effective strategy to entice the employee into sharing their knowledge.
3. Monitoring is really hard so rely on performance management, goals and objectives rather than monitoring. This also shows the trust that you have in your employee, and this also builds trust in the organisation which will even further boost employees to share their knowledge.
4. Make sharing easy.
5. Change your physical space a bit in the organisation to allow more collaboration. It might lead to forming more communities of practices (CoPs) which may help the organisation in sharing more knowledge.

We should also consider customer feedback as this may help in finding the solutions of some problems and also in influencing future design.

When such practices are followed, employees would indeed share knowledge, and when we give them incentive for their hard work, they give more positive results.

Accessing: Once the knowledge sources are identified, they are then collated into background references for a repository and in order to facilitate access and retrieval for all the people in the organisation. Organisations use focus group to arrive at a consensus as to how best this can be achieved. Since this knowledge is shared with knowledgeable people, the process of accessing will address difficult problems, and we get a second opinion from the experts about this knowledge which has been coded and shared. One can access and retrieve knowledge directly from repository as well, e.g. using a system of knowledge base to obtain an advice on how to read a knowledge document or how to do something in order to arrive at a decision.

Fig. 2.6 Framework to
show how knowledge
conversion happens in KM
cycle (source: Nonaka and
Takeuchi)

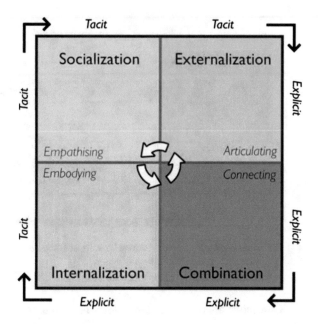

Application: Knowledge application refers to the actual use of knowledge which has been created and captured and put into the KM cycle. When capturing, coding, sharing, etc. are done, then we are still in the third quadrant of Nonaka and Takeuchi model as shown in Fig. 2.6. To complete the knowledge spiral, we must do internalisation of knowledge and this happens in application part of KM cycle.

Knowledge has a short shelf life sometimes, so we should use it as quickly in the organisation as possible because that thing which you have shared may become outdated and something better might have come in the market so using it quickly is not such a bad option.

This step in KM cycle is the most important one because if this step is not accomplished successfully then all of the KM efforts will be in vain. Remember that KM can succeed if knowledge is used.

One of the ways by which knowledge application is done at individual level is by task analysis and modelling. Task analysis is the process of describing a task by breaking it down into its primary components. It is done by task decomposition. For example, a process of polishing your shoes is described below:

1. Wipe off the dust off the shoes by using a cloth or a brush.
2. Apply the polish on one shoe and then leave it aside for the shoe to absorb the polish.
3. Take the second shoe and apply polish on it and leave this aside.
4. Now take a brush and rub it on the first shoe vigorously for a couple of minutes.
5. Repeat step 4 for the second shoe.
6. You are ready now to wear the shoes.

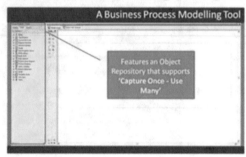

Fig. 2.7 How knowledge capture is done by software (source: Synoptic Consulting Sydney)

Knowledge application at group level is done in the following ways:

1. *Knowledge Reuse:* It involves recall, recognition and applying the knowledge. There are three major roles required for knowledge reuse: knowledge producer (the one who produced the knowledge or documented it), knowledge intermediary (the one who prepares the knowledge by indexing, standardizing, categorization, mapping, publishing etc.) and knowledge reuser (the one who understands, retrieves and applies it).
2. *Knowledge Repositories:* An example of this is already shown in Fig. 2.7. Knowledge repository is nothing but a place to store and retrieve the knowledge. It can contain a mix of tacit and explicit knowledge. They are usually intranets or portals. They serve to preserve, manage and build up the organisational memory.

The tools for knowledge application are:

1. *Adaptive Technologies:* These technologies are used to better target the content to a specific knowledge worker or to a group of knowledge workers who share a common work need. If a worker wants that his content should be in a particular language or a particular format, then he gets that in these technologies.
2. *Intelligent Filtering Tools:* The name itself suggests what they are used for. These tools can help in addressing the challenge of information overload by just selecting relevant content and delivering this in a just-in-time and just-enough format as required by the user.

2.3   KM Cycle Transformation From Tacit to Explicit and Back to Tacit

Explicit knowledge is that which can be expressed in language, can be easily articulated, can be understood, can be codified and can be recorded, whereas tacit knowledge is one which cannot be expressed in language; it is unseen knowledge. It cannot be easily codified. Tacit knowledge is habits, assumptions, skills, etc. Explicit knowledge said to be only the tip of the iceberg. Our aim should always be to convert tacit knowledge to explicit knowledge. Nonaka and Takeuchi say that tacit and explicit are part of a four-part process. Some people said earlier that converting tacit knowledge to explicit knowledge is impossible. But Fig. 2.6 shows how KM cycle undergoes the knowledge conversion process in an organisation. The four modes of knowledge conversion are:

Socialisation: The word socialisation itself speaks a lot. It is nothing but sharing knowledge through social interactions. It happens when tacit knowledge is converted to tacit knowledge. Here take an example of a person X who is having 20 years of work experience in an organisation. When this person meets a fresher Y during a coffee meeting at the canteen and gives the tacit knowledge that he has to this newcomer, then this is known as socialisation. Here knowledge is created long back in the mind of X, and this knowledge is still not written down when Y is capturing it in his brain.

Advantage:
This is the easiest form of transferring knowledge.
Disadvantage:

1. The knowledge still remains tacit and our aim should always be to make it explicit.
2. Honda does brainstorming camps for solving difficult problems. This is one of the examples of socialisation.
3. It is very time-consuming and difficult to use this mode.

Socialisation happens through:

1. Interaction.
2. Teamwork and a culture of sharing one's experiences.
3. Informal communication.
4. If in an organisation, workplace has open space without any physical barriers, then a lot of tacit to tacit to transformation happens.
5. Other examples being corridor meetings, apprenticeship, mentoring models, knowledge days, knowledge camps, etc.

Externalisation: It happens when tacit knowledge is converted to explicit knowledge. When knowledge is in written form, then we can read it, and if we read information it makes it understandable and interpretable. In this mode, individuals will be able to understand things. In this mode, the previously tacit knowledge can be written down, recorded, taped, etc., and for this we need intermediaries such as knowledge journalists.

Externalisation is done through:

1. Writing notes
2. Brainstorming
3. Encouraging a learning environment

Combination: It happens when explicit knowledge is converted to explicit knowledge. In this process no new knowledge is created, but the knowledge is organised into broader concept systems. Knowledge is logically ordered to make it more meaningful. Here categorising, sorting, reconfiguring, adding and updating content are done. These systems can include:

1. Databases
2. Books

A working example is formal education and training.

Internalisation: It happens when explicit knowledge is converted to tacit knowledge. The understanding of explicit knowledge is internalisation. It transforms to tacit and becomes part of an individual's basic information. In a nutshell, it is nothing but learning by doing. Examples of internalisation are:

1. By practising and repeating
2. By experience and expertise

This creates know-how. For example if you learn to ride a bike, you never forget it throughout your lifetime.

The advantage of SECI (socialisation, externalisation, combination, internalisation) model is quite simple and proves that tacit knowledge can be converted to explicit knowledge in an organisation.

Disadvantages of SECI model:

1. Based on the Japanese companies, where employees often have 'jobs for life'
2. Heavier reliance on tacit knowledge in this model

So we can come to a conclusion that SECI model shows that tacit knowledge can be converted to explicit. It is a framework for managing KM processes. These processes are:

1. Socialisation—verbal explanation
2. Externalisation—physical demonstration
3. Combination—reading the manual
4. Internalisation—embodying knowledge

2.4   Types of Knowledge Management Cycle

There are four types of knowledge management cycle. They are:

1. Zack KM cycle (1996)
2. Bukowitz and Williams KM cycle (2003)
3. McElroy KM cycle (1999)
4. Wiig KM cycle (1993)

In the different KM cycles, the term used differs but the real difference is not much.

2.4.1   Zack KM Cycle

The physical products follow within an organisation and can be applied to the management of the knowledge assets. Although the Zack KM cycle provides information about physical product, it can be easily extended to the knowledge products.

It is composed of the technologies, facilities and processes for manufacturing products and services. Their approach provided a number of good analogies like the notion of a product platform (the knowledge repository) and the information process platform (the knowledge refinery). The repository becomes the foundation of the company to create a family and knowledge products. The KM cycle basically means creating higher-value products from the existing sets of knowledge. In Zack's approach, the interfaces between each stage are designed to be seamless and standardised.

Mayer and Wack analysed their major development stages of the knowledge repository and mapped following stages on to a KM cycle:

1. Acquisition
2. Refinement
3. Storage
4. Distribution
5. Presentation (Fig. 2.8)

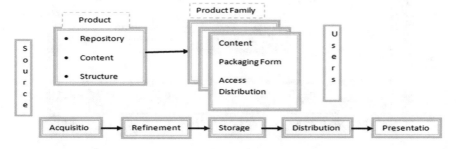

Figure 2.8 Zack KM cycle

Meyer and Zack KM cycle stages	Significance
Acquisition	• Quality control of the data
	• The data is measured on parameters like depth, relevance, credibility, cost, control and exclusivity
Refinement	• Primary source of value addition to the primary data
	• The value addition could be physical or logical
	• It also means to standardise the primary data by cleaning up the irrelevant materials
Storage	• It forms a bridge between the upstream acquisition and refinement stages that feed the repository and product generation
	• It could be physical or digital
Distribution	• The process to deliver the product to the end users
	• It encompasses not only delivery channel but also its timing, frequency, form and language
Presentation	• This is the cumulative effect of each and every stage of the KM cycle
	• If it has been able to create value here, it has been able to find the right usage; then the KM cycle has been successful

2.4.2   The Bukowitz and Williams KM Cycle

It describes a knowledge management framework which helps the organisation to take strategically correct steps for the creation of knowledge in the organisation.

In this framework, knowledge consists of knowledge repositories, relationships, information technologies, communication networks and organisational intelligence (Fig. 2.9).

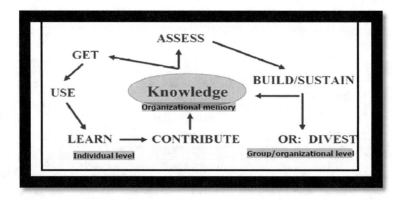

Fig. 2.9   Bukowitz and William KM cycle

Bukowitz and William KM cycle stages	Significance
Get	• This is the first stage that consists of seeking out information needed to make decisions and solve problems
	• The main problem is obtaining tacit knowledge
Use	• This includes fostering the info obtained for dealing new problems in the organisation or for innovation
Learn	• The use of knowledge can result in failure or success. This needs to be reflected
	• Without it the knowledge will be of no real significance for further use
Contribute	• The employers post what they have learnt to the communal knowledge base
Assess	• It is done for intellectual capital
	• The present capacity is measured with future needs
Build	• It ensures that the future competitiveness of the organisation is maintained and sustained
Divest	• This is a let go stage wherein if the organisation can better use their intellectual capital externally, it should have means for it
	• Similarly the cost/benefit of holding and divesting the info is considered

Strengths: Introduction of two new critical phases:

1. Knowledge content and the decision as to whether to maintain or divest this knowledge
2. Consider both tacit and explicit knowledge (more comprehensive than the Meyer and Zack cycle)

2.4.3   The McElroy KM Cycle

The McElroy emphasises that the organisational knowledge remains in the minds of the individuals and groups and objectively in the explicit terms. He describes the KM cycle that consists of the processes of knowledge production and knowledge integration, with a series of feedback loops to organisational memory, beliefs, claims and business process environment (Fig. 2.10).

Definition of single-loop learning and double-loop learning:

In *single-loop learning*, individuals, groups or organisations modify their actions according to the difference between expected and obtained outcomes.

In *double-loop learning*, the entities (individuals, groups or organisation) question the values, assumptions and policies that led to the actions in the first place; if they are able to view and modify those, then second-order or double-loop learning has taken.

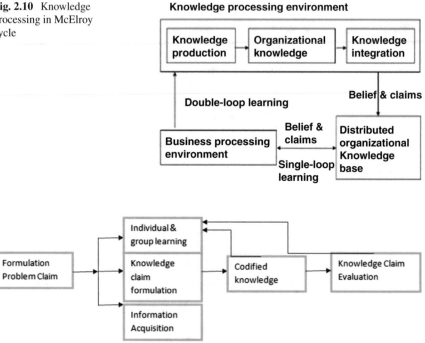

Fig. 2.10 Knowledge
processing in McElroy
cycle

Fig. 2.11 McElroy KM cycle

The following diagram depicts the main stages of the McElroy KM cycle
(Fig. 2.11):

McElroy KM cycle stages	Significance
Formulation problem claim	• An attempt to learn and detect the specific gap in the knowledge asset
Individual and group learning	• It is the validation of the collected knowledge by the organisation
Knowledge claim validation	• Involves codification at an organisational level based on individual and group innovations
Information acquisition	• The process by which an organisation acquires knowledge claims or information produced by others, usually external to the organisation (competitive intelligence)
Knowledge claim evaluation	• The process by which knowledge claims are evaluated to determine their validity and value. The knowledge is of greater significance than the present knowledge

2.4.4 Wiig KM Cycle

Wiig basically focuses on the three conditions necessary for the organisation business conduction:

1. It must have business and customers.
2. It must have resources to satisfy the need of customers.
3. It must have the ability to act.

The third point has been emphasised in the Wiig KM cycle. Wiig emphasised on the smarter execution for achieving better productivity by applying proper knowledge (Figs. 2.12 and 2.13).

Significance:

This model provides a 'clear and detailed description of how organisational memory is put to use in order to generate value for individuals, groups and the organisation itself'.

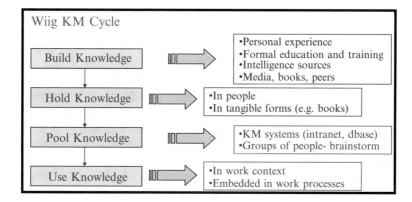

Fig. 2.12 Wiig KM cycle

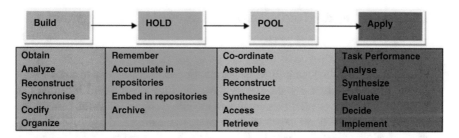

Fig. 2.13 Summary of Wiig KM cycle

2.5   Knowledge Management: Perspective of Small and Big Organisations

2.5.1   Small Organisations

It is easier for the smaller organisations to emulate individual knowledge management cycle since the number of employees is less and the level of interaction is high. But nevertheless they need to inculcate the habit of KM right from the beginning to ensure that the knowledge does not get lost in the day-to-day operations of the organisation. There could be resource constraints in the earlier days of functioning of an organisation, yet the background of KM could be managed. But the incentives to inculcate KM should be clear in the minds of the managers. The result of not capturing knowledge in the earlier days of an organisation could result in loss of competition and incurring extra cost to obtain the knowledge. The smaller organisations are flexible enough to tap the changes so in a way they are at an advantageous position in comparison to bigger organisation.

2.5.2   Large Organisations

As the organisation grows, it becomes difficult to emulate KM cycle of an individual. A lot of stakeholders raise queries and the need to share information rises. The dissemination of the knowledge becomes difficult due to time and space crunch. But the bigger organisations have the resources available, and with the advent of modern technology, it is possible for larger organisation to implement KM in their organisation. More than anything, these large organisations need to consider a shift in their culture, if it is not tuned to the implementation of the KM. They need to set their priorities to implement KM since its need can be latent, and by the time it could become urgent, it would be late and there could be serious loss of competition.

2.6   Challenges for Implementing Knowledge Management Cycle

- *Challenge 1:* Moving people from information-driven process to knowledge-driven process—Most people donors have the time energy and incentives to invest in the creation and dissemination of the knowledge. Unless the benefits are not obvious, it is difficult to get the work done. In creating KM cycle, the managers face similar issues where he has to signify the importance of KM to the employees.
- *Challenge 2:* The capturing of tacit knowledge—No process is full proof, and the capturing of tacit knowledge remains a major challenge for any KM cycle.

- *Challenge 3:* To delink technology and KM—The use of technology has created myth among the organisations that KM is meant for big organisations which can spend on the technology. But technology unavoidable part of any modern Knowledge Management practice.
- *Challenge 4:* To create trust for free flow of information—The cultural issues of the organisation affects the type of KM cycle that can be implemented. In a close organisation the implementation of KM would be difficult and will the need necessary changes in the culture.
- *Challenge 5:* The implementation of KM cycle will change not only with the change in the organisations but also due to the changes within the organisation itself.

2.7 Summary

This chapter discusses different stages in knowledge management cycles. Each KM cycle has stages such as capturing the knowledge, codifying the captured knowledge (this will help in easy retrieval later on), getting it reviewed and then publishing it. However, in the knowledge economy, the effectiveness of KM cycles is measured by the degree to which the knowledge is shared among employees. Hence, without sharing and applying these knowledge to business practices, the KM cycle can never be completed. In the modern times, these knowledge are 'pushed' to potential business users. From tacit to explicit, the change in type of knowledge is important for business to be productive. This change is obtained by following a systematic and scientific way of adopting a KM cycle.

There are different KM cycles that have been proposed by different experts. However, an organisation needs to adopt different stages to suit its requirements. In the process of adopting these KM cycles, there will be hurdles that the organisation will have to overcome. A successful learning organisation will be able to overcome these challenges and use knowledge as a strategic tool.

Chapter 3
Communities of Practice for Effective Knowledge Management Strategy

3.1 Understanding Community of Practice

'Community of practice' as a term is of fairly recent origin; however, the basic phenomenon that it refers to is far older. This concept provides a useful perspective on learning and knowing. Thus an ever-increasing number of people and organisations have started using this very practice to enhance knowledge and know-how, within their organisational framework.

On a very basic level, and for the layman, a community of practice may include a group of people who share an interest, a craft and/or a profession. A natural evolution happens within the group as a result of the commonalities of interest in one particular area, sector or domain. In some cases, it is very goal oriented, with the formation of the CoP directly related to the actual target of attaining knowledge pertinent to the field of interest. Thus the participants make their learnings from each other through sharing of information, knowledge and personal experience and thus have a chance to develop their personal and professional goals.

The knowledge held by a company can be roughly divided into four types. The three most relevant to communities are described below (Fig. 3.1).

3.1.1 Strategic Competence

These areas include new and evolving knowledge and information which will be important in the future, but on which, in the present situation, a lot is yet to be learnt. It is then managed by the implementation and the set-up of certain programmes related to acquisition of knowledge such as R&D projects, active learning projects or, in the case of truly multinational companies, through groups related to global

© Springer International Publishing Switzerland 2016
S. Mohapatra et al., *Designing Knowledge Management-Enabled Business Strategies*, Management for Professionals, DOI 10.1007/978-3-319-33894-1_3

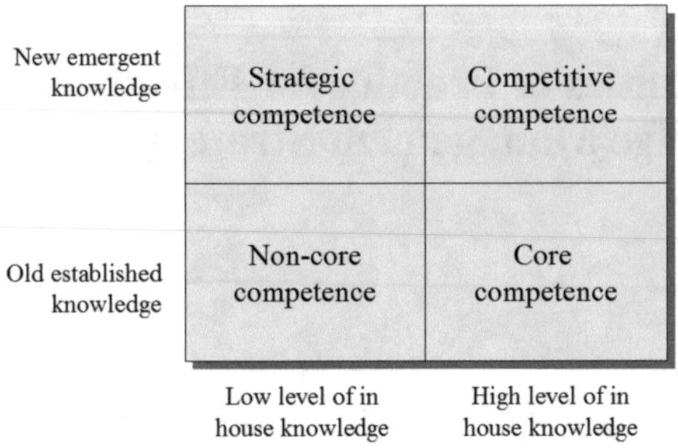

New emergent knowledge	Strategic competence	Competitive competence
Old established knowledge	Non-core competence	Core competence
	Low level of in house knowledge	High level of in house knowledge

Fig. 3.1 The areas of organisation related to knowledge

best practices. Thus the focus is on acquiring knowledge from start-up/pilot projects and programmes. As the level of knowledge increases, we come to the next level, that of competitive competence.

3.1.2 Competitive Competence

The main area now is of again new and forward moving knowledge, but in this case, the organisation is already aware and conscious about it. Thus a competitive advantage is acquired. To outperform its rivals, the organisation must learn the fastest and learn the best. As the holders of this knowledge number are many, it is well circulated among different levels of the organisation. However the process of capture of knowledge must be a continuous one, enabling CoPs and gradual improvement leading to establishment of best practices. After the establishment of best practices procedures, the knowledge will find its way to major application in the business and become well regarded and well reputed. Then the next level will be reached, that of core competence.

3.1.3 Core Competence

These are areas of recognised knowledge and well-established sources that the company is very well acquainted with. It is likely to be at the core of all their deliveries and is generally the crux of all business activities of the organisation. Needless to say, efficient and effective management of these areas are essential for the organisation to perform. Thus after the consolidation of knowledge, the evolution of certain standards is now important. At this stage, the focus shifts to the core competence and best practice, which becomes a part of all practices of work and all different procedures pertaining to the organisation. Thus the management of knowledge focus now lies on making it standard across the company, and these very standards are constantly reviewed and also updated. Thus all technical functions and experts work closely with and within communities of users in the organisation (CoPs).

3.2 Characteristics of Communities of Practice

Therefore we see that CoPs are in general formed by people who are associated with a process of combined and joint learning, in a joint area or field: tribes surviving in the wild, musicians working on new sounds, technicians seeking to find common solutions, scientists looking for better research and doctors seeking better practices of surgery and treating patients. However a distinction must be made between 'community' and 'communities of practice'. There are certain characteristics which are absolutely crucial for CoPs (Fig. 3.2):

The Domain: A CoP must be distinguished from a group of friends or any network with various connections between people. There must exist an identity which has to be definite with a common field of interest. Membership to such a group in turn will imply a certain commitment to that very field or area of interest; it is this very commitment to a shared capability that differentiates members of the CoP from other people. It is not necessary for the area of interest to be recognised as technical expertise or know-how by outsiders. A group of mothers caring for their children may have a certain identity and deal with their domain, i.e. looking after their babies. They will value their shared and collective capacity and learn from each other, though this may not be valued or recognised outside of their particular group.

The Community: In the pursuit of their area of interest, the group members may engage in joint discussions and other activities to share knowledge and help one another. There is a certain building of relationships which enable them to understand, ascertain and learn from one another. They do not necessarily have to work together on a daily basis, but the meetings do occur regularly, even if the activity they may be engaged in is done alone; the meetings help to discuss problems and come up with solutions and uniform procedures.

The Practice: A CoP is not just an interest group—people who follow only certain kinds of movies, for example. The actual members of a COP are essentially practitioners. They have a shared practice and engage in developing a range of resources which may include personal experience, efforts in managing problems and certain tools. However, this takes an invariable amount of time and continuous interaction over a period. An interesting talk may give one particular insight but it does not become a CoP. Shared practice may be unconscious but may also be a self-assertive development. Surgeons who may meet at the doctor's lounge may not realise that during their talk, they may know more about the different patients at a clinic. Thus with the help of these conversations, they may have developed a certain collection of personal recollections and cases that becomes part of the range of their practice.

With a combination of these elements, a successful CoP is established (Figs. 3.3, 3.4, 3.5, 3.6 and 3.7).

Fig. 3.2 Characteristics of community of practice

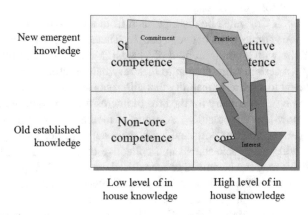

Fig. 3.3 The range of interest and knowledge within an organisation

3.3   Four Types of Community

For the purpose of better explanation within an organisation, four main types of communities are chosen. They will work in different ways, will be of variable sizes, will certainly have varied range of application and will tackle and address information and sharing of knowledge activities with different levels of responsibility. The organisation will need to identify its type:

Innovation Community

- Chaotic pool of thinking individuals
- Pooling together varied knowledge to come up with novel innovations
- Small convergence of in-house experts—with assigned roles
- Action learning group
- Dealing with comparatively lower levels of in-house capacity

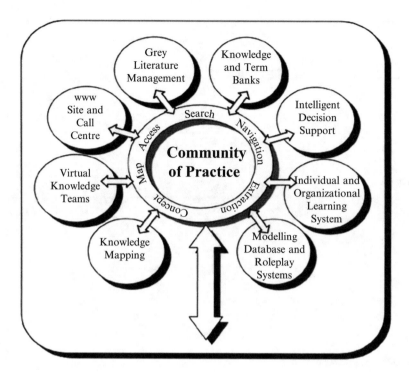

Fig. 3.4   An overview of community of practice

Fig. 3.5   Degree of identity

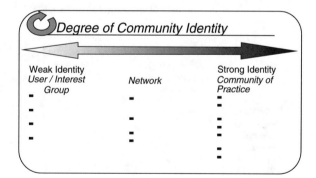

- Seeking to create resolute in-house capability

Exploring Community

- Parallel learning styles
- Continuous exchange of experience
- Tacit knowledge
- Major emphasis on dialogue and discussions
- Voluntary or assigned learners

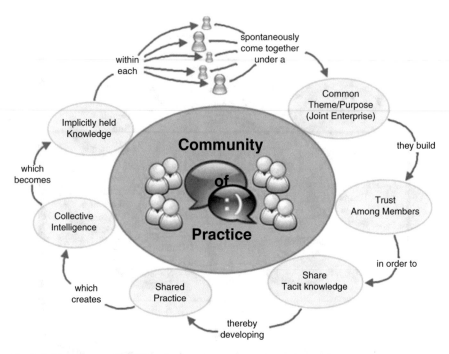

Fig. 3.6 The process of COP learning

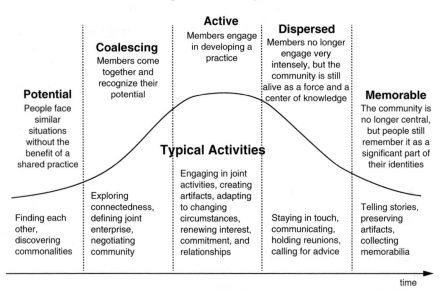

Fig. 3.7 Stages of development

- Takes into consideration the rapid evolution of knowledge with low capacity levels
- Developing effective practice

Best Practice Community

- Approaches to different scenarios are compared.
- The best is chosen.
- Emphasis on building of KM assets.
- Certain standards are worked upon.
- Absolute core of experts in fields.
- Wide range of members-users.
- Reasonably well-known practices and knowledge are worked on.
- Seeks to define and bring into practice the best methods.

Stewardship Community

- Standard maintenance.
- Updates as and when required.
- Monitoring and evaluation of standards and their practicality.
- Teaching, training and mentoring.
- Small number of experts.
- Wide range of members-users.
- Well-established knowledge is worked upon.
- Maintaining a certain fixed standard for operations.

3.4 The TATA Steel Experience

Knowledge Management and Communities at TATA Steel

TATA Steel launched its KM initiative when a number of recurrent and similar breakdowns in its manufacturing facilities were traced back to the lack of a culture and systems within the organisation, to tap into its own expertise and experiences. The company entered into the new millennium '...With a confidence of a learning and knowledge based organisation'.

Their KM programme was '...To capture the available abundant knowledge assets in the form of tacit (experience, thumb rules, etc.) and explicit (literature, reports, failure analysis, etc.), to organise and transform the captured knowledge, and to facilitate its usage at the right place and in right time'. Apart from investing in IT infrastructure, the company incorporated participation in KM activities in staff members' performance appraisals. Knowledge communities were created for its core processes and systems—there are about 20 communities functioning now (for about 4000 executives).

(continued)

(continued)

The communities are a knowledge-sharing platform, not a task force to solve a problem—they bring people together to share what they know and to learn from one another. Since many of these are large in size, there are further specialised sub-communities who federate into that community.

Each of the knowledge communities has a few distinct roles viz., Champion, Convener, Practice Leader, Lead Expert and Practitioners, etc. Communities meet face to face every month, and hold presentations, discussions, visits, sharing of experiences, talks by experts, etc. Some salient features of the company's KM approach were:

- *Strong support and commitment of Senior Management to KM.*
- *Performance measurement system that takes account not just of tasks performed and roles fulfilled, but also a KM index.*
- *The approach has been of documenting; simplifying, refining and then passing on knowledge for use using pedagogically simple methods (in preference to loading personnel with dense material)—a combination of 'hard' and 'soft' knowledge.*

Source: www.TATASTEEL.com

3.5   Virtual Communities of Practice

A virtual community of practice (VCoP) can be defined as a network of people/individuals who are a part of an area of interest, about which they mainly communicate and share thoughts on, online. These practitioners share problems and solutions, methodologies, experiences and tools, and all such communication contributes to the development and accumulation of knowledge within the CoP and improves the knowledge of every individual practice. A point to note is that VCoPs are highly unlikely to work on joint projects together but will in most cases share news and also advise on areas of professional or academic nature and interest.

In a VCOP, a more informal process of learning occurs and involves a process of participation in a sociocultural practice; thus in a way, more experienced practitioners pass on knowledge and skills to newcomers or even peers, which they have acquired, and consequently, the collective expertise of all participants increases. With technological developments, academicians across the world have a chance to work in much improved environments of learning through increased communication, interaction and incorporation of pan-national models which have all been possible due to information communication technologies.

3.5.1 Barriers

A potential first barrier may be the *actual discipline* that is involved. In some areas and domains such as the sciences, new and absolutely cutting-edge knowledge may be extremely hard to pass on through a virtual channel to large groups. Open communication is often more difficult with the culture of freedom and degrees of independence that is currently enjoyed by academics. Sharing of tacit and transactive knowledge is often easier to share with personal and close contacts.

Another barrier is of course the *selection of the information communication technology*. Almost all professionals and academics or students use ICTs to match their usage to match their operational needs, irrespective of their skill levels in the use of the same. Thus this brings the important issue, whether the COP is task or practice oriented. A task-based VCOP may be short-lived in order to attain a specific activity, and a subject/practiced based VCOP may develop slowly and be less transient.

The final issue may involve the *use to ICTs to bridge geographical divides* which can lead to misinterpretation of messages due to the absence of non-verbal cues which may be missing from communication. The actual face-to-face communication richness will be missing. Thus much information may be misinterpreted online without the instant cues and feedback of a person-to-person session.

Factors for Success of VCOPs

1. *Technology and its usability:* VCOPs need to make pertinent use of internet technologies. A VCOP at its initiation would have to ensure that all participants have the technical provisions and skills in order to facilitate mutual engagement.
2. *Communication:* This is another critical success factor (CSF) for the development of an environment based on mutual trust and understanding within the community, in order to enable the growth of the COP and also to enable it to change and achieve its objectives. Continued Interaction helps develop common values which again leads to a healthier environment.
3. *Membership:* This is identifying group members with a prior knowledge of each other and paying attention to cross-national and cross-cultural dimensions in international online communities which adds to the complexities and challenges but ultimately adds value to the COP. This is very well illustrated by the HSBC advertisements which identify different gestures which convey various contradictory meanings in different cultures.
4. *A sense of purpose:* This is essential for the COP to actually achieve its goals, especially in a task-oriented scenario. An offshoot is the role of leadership, as the community becomes more diversified and distributed; it is essential for good-driving leadership, thus following the proper use of netiquette and the way guidelines on good practice are put into practice by facilitators. Any inflammatory behaviour within the community may inhibit contributions from neo-members, also user-friendly language and graceful ways of bringing people into conversations.

Benefits, Barriers and Critical Success Factors

The following table lists the various benefits, barriers and the CSFs for successful VCOPs:

Benefits	Barriers	CSFs
• Enhanced learning environment • Synergies created • Capabilities extended to higher level • Knowledge sharing & learning • Gaining insights from each other • Deepening of knowledge, innovation & expertise • Cyclical, fluid knowledge development • Feeling of connection • Ongoing interactions	• Perpetuation vs. Change and diversity • Disciplinary differences • Culture of independence • Tacit knowledge • Transactive knowledge • Specialist language • Collegiality, strong physical community • Shifting membership, Creating and maintaining information flow • No F2F to break the ice • Read-only participants (formerly lurkers) • Hidden identities	• Good use of Internet standard technologies • Technological provision • ICT skills • Institutional acceptance of ICTs as communication media • Good communications • Trust • Shared understanding • Prior knowledge of membership • Sense of belonging • Cultural awareness • Sense of purpose • sociocultural practices • Neo-apprenticeship style of learning • Practice-based usage adopted personas • Selectivity in ICT use • Task-based usage • Sensitivity in monitoring, regulating, facilitating • Netiquette • User-friendly language • Time to build up the CoP • Regular interaction

3.6 Development and Nurturing of CoPs

Sometimes communities of practice arise naturally, but this does not mean that organisations can't do anything to influence their development. Most communities of practice exist whether or not the organisation recognises them. Many are best left alone—some might actually wither under any institutional spotlight. And some may actually need to be carefully seeded and nurtured. But a good number will benefit from some attention, as long as this attention does not smother their self-organising drive.

There are a few discrete roles to be played in the community, the key role being the community leader, coordinator or facilitator—the person who is accountable for ensuring the community functions as a knowledge-sharing mechanism and best practice is identified and shared. This person is involved in the start-up and growth of the community and in developing and maintaining the community processes. The choice of a good leader is crucial to the effective operation of the community. This recognised networker within the community facilitates the linkages and relationships between the members, as well as potentially stewarding any community output. A good coordinator leads from within, energises the community and builds a feeling of trust and ownership among the community members. The main tasks of the leader include:

- Build membership.
- Maintain activity and energy.
- Set the behavioural style of the community.
- Manage discussions.
- Manage relationships/contact brokering.
- Liaise with the sponsor.

Whether these communities arise spontaneously or come together through seeding and nurturing, their development ultimately depends on internal leadership. Certainly, in order to legitimise the community as a place for sharing and creating knowledge, recognised experts need to be involved in some way, even if they don't do much of the work. But internal leadership is more diverse and distributed. It can take many forms (Fig. 3.8).

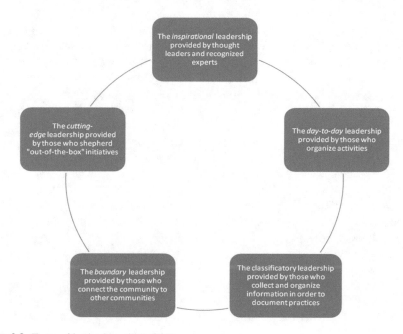

Fig. 3.8 Types of leadership within COPs

3.7 Communities of Practice in the Development Sector

Though the basic idea behind the COP movement is the same in the development sector, as seen in the case above, some salient features need to be looked into in order to make it more successful and wide reaching. Some basic tenets must be followed when implementing COPs in the development sector; these may include:

- Make sure key **stakeholders** are members.
- Be aware of the specificity of the subject (domain).
- Care for shared internal rules, a code of conduct.
- Keep the energy flowing.
- Create links between (different) stakeholders and their realities.
- Adjust to changes in the environment.
- Strive for most practical and tangible outputs/outcomes.
- Stay aware of ownership.
- Make the resources available.
- Select carefully the ways of communication, of 'being connected'.
- Focus on the value of the COP for the members.

Adikke Pathrike: **Farmers' Journal Builds a CoP?**
The medium-to-heavy-rainfall districts of southern, coastal Karnataka and northern Kerala, are characterised by low hills and lateritic soils, making the area unsuitable for the intensive rice production. Farmers in the area grow small amounts of rice and other food crops but most of their efforts are devoted to cash crops like arecanut, coconut, cashew, cocoa, and black pepper. A particularly severe drop in prices in the mid-1980s caused a crisis in Karnataka. Several committees were formed through the association and Shree Padre, a local journalist (and areca grower), volunteered to produce a newsletter for growers, on an experimental basis. The positive response to the newsletter led to a search for a more extensive magazine: and Adikke Pathrike was born. A monthly magazine, about 28 pages is attractively produced, with a colour photograph on the cover, additional black-and-white photos and diagrams to illustrate news items. It is written entirely in Kannada, the local language. The magazine is 15 years old now, has never missed a deadline, and brings out a special annual issue. Beginning with Areca, the publication soon expanded to discuss a wide range of crops and other rural activities: management techniques for crops, prospects of new crop enterprises, farm machinery, farm household improvement, and even new recipes. The editorial stance of the magazine is 'pragmatically green'—favouring technologies that lower dependence on external inputs. Furthermore, advertising from pesticide manufacturers is not accepted. The magazine is very 'science friendly' and welcomes information on new technology. Some of the key features are:

(continued)

(continued)

- *It tries to ensure that farmers themselves write about their own experience, rather than simply passing on information from university or public research institute scientists.*
- *An iterative and adaptive approach to technology description, i.e., in many cases an initial description of an innovation or new technique by one farmer may be amended, elaborated, or challenged by other farmers' experiences in subsequent issues.*
- *To encourage adaptation, there is a consciously strategised promotion of farmer-to-farmer contact. This is achieved through a question and- answer section in which farmers share their experiences and ask for advice. Also, each article provides the farmer author's contact details.*

A sceptical and investigative stance towards newly promoted crops or technologies, particularly those featured in positive terms in the conventional press. Unlike the normal creation of a community, here there has been a product— Adikke Pathrike—that bought itself or attracted a community of readers and contributors. This was also facilitated through the journalism workshops that created a team of potential contributors. Members are mostly farmers in the area and there is a core of 30–40 'friends' (workshop graduates including farmers) who visit farmers, and work with them to produce an article. The author is paid a small honorarium. In other cases, farmers themselves write brief communications or at least inform Adikke Pathrike of an interesting story. The articles are always based on first-hand farmer experiences although these may require significant rewriting and editing. The combination of the cover price and advertising revenue supports a staff of five people. The magazine is a non-profit entity, registered as a trust. Adikke Pathrike has demonstrated a significant demand for the opportunity to share experience, to stimulate experimentation, and to take a broader view of the farm household. An important part of its success is that innovations are discussed alongside more mundane issues (such as controlling houseflies) that contribute in an important way to the quality of rural life and to the incentives for pursuing technological change. An offshoot of the exchange in Adikke Pathrike is the establishment of a seed exchange group that meets monthly to share experiences and exchange seeds. The group is now 10 years old and has a membership of 30–40 farmers.

Source: CAAM Website (www.farmedia.org/index.html) and records of telecommunication with Shree Padr

3.8 Summary

COP is the starting point for knowledge creation, where a knowledge is tried out, reviewed and vetted by so-called community of experts. The COP, in olden days, used to be formed out of personal traits, friendship and acquaintances. However, in modern knowledge economy, a business has to achieve its business goals, and hence the COP is formed from business point of view. As a result, the community of experts are grouped together by business objectives rather than personal acquaintances. This chapter deals with the approach to form a COP for meeting business requirements. The chapter discusses different factors that will lead to be a successful COP. A special section has been earmarked for development sector. This section discusses the need for COP in NPOs and NGOs and how to make these COPs effective.

Chapter 4
Issues and Challenges in Knowledge Management

4.1 Introduction

Knowledge-based economy has taken it over from the money-based economy. Huge exchanges of know-how and advanced technology have resulted in transforming the picture of many developing countries. Economic development has always centred on knowledge being acquired by human beings which later came to be known as 'human capital'. But only over the last few years has its relative importance been recognised, just as that importance is growing. This has led to change in economic functions, and economists have already started putting it as an important variant in all laws that govern economics. With growth of information technology and people realising the power of knowledge, companies can no longer ignore the importance of the accumulated knowledge they have at their end.

As access to information becomes easier and less expensive, the way in which we use the knowledge and what skills are being developed based on this knowledge becomes noteworthy. While working on a particular problem, we need two types of knowledge: one is codified knowledge which is more or less like a tool. The other type of knowledge is tacit knowledge which tells us how to use the tool. Moreover the trend these days in knowledge is more towards recognising relevant information and patterns of information and interpreting it.

To understand what is restricting the growth of knowledge network and how to use it in an optimum way is to enable the growth of world economy and also to see that we also look to train people through this to make them worthy for the challenges of the future. In order to do so, we have to look at the currents flaws and identify the loopholes of the system. This particular paper tries to bring together some of these issues and discuss it at length.

© Springer International Publishing Switzerland 2016
S. Mohapatra et al., *Designing Knowledge Management-Enabled Business Strategies*, Management for Professionals, DOI 10.1007/978-3-319-33894-1_4

4.2   Challenges Faced by KM in General

1. *Addressing people issues*: The people of an organisation form an essential part of a KM strategy being implemented within the organisation. Due to the diversity of the workforce, it becomes difficult to bring the people onto the same stage as everyone. Due to the disparity among the individuals, the company may miss out on the mission statement. Similarly the vision statement is not properly shared with the team; there may be lapses in use and also misuse of the knowledge management systems. A new entry into the organisation and an exit of an old employee change the KM base of an organisation. It is also important that the new employee gets integrated into the system seamlessly and the old employee gives the maximum of his owned knowledge.

2. *Employee retention, development and assimilation*: Employees of any organisation are its biggest assets. So any type of new development which comes up has to be done by taking these employees into consideration. But with the growing attrition rates in all industries and shifting of knowledge base from one camp to another in some ways, employees are turning out into liabilities for the company. So to keep the KM up and growing, it will always be essential to keep the workforce intact and make them realise the importance of assimilating this type of knowledge and encouraging them to make use of such knowledge whenever possible.

3. *Information overload*: As the knowledge management process ages and matures over time, the knowledge keeps on accumulating and building up. Proper codification techniques can be used to get to the exact data in low amount of time. But with huge amounts of data, maintaining relevancy of the knowledge becomes a big question. So with huge amounts of information, dating back hampers creation of an environment rich for organisational learning, innovation and creativity.

4. *Collaborating challenge*: Collaborating between different organisations and departments is very important these days as most of the knowledge is interdisciplinary. A set of data standing alone doesn't mean a thing until it is combined with different scenarios. In order to get the bird's-eye view of the picture, it is always essential to get together all your stakeholders together. Similarly in case of KM, there are a lot of other partners who work towards completing the knowledge structure. Whenever we are looking towards the growth, any organisation has to grow keeping its well wishers in tow.

5. *Conversion of tacit to explicit knowledge*: Tacit knowledge remains in the subconscious minds of the people. There are times when it may exist but the person is unaware of it. So the conversion has to be done at the same time during which it is being practiced. But generally companies postpone this process till the point the person is leaving the company. During the last few days that he has with the organisation, he is expected to transfer his knowledge or convert it to explicit knowledge which rarely happens correctly.

6. *Building a process-driven system*: The process-driven system aims at reducing the human errors. KM solutions reduce cycle time errors which are associated with decision-making, problem-solving, documentation and appropriate capture which help the organisations to become more successful and profitable. This greatly decreases rework and increases efficiency. But the major challenge is to make the age-old organisation work like clockwork. Perfectly process-driven companies exist only on paper, and knowledge management includes so much of human interface that it can hardly be entirely process driven.

7. *Effective use of KM*: The efficiency of a KM is judged from the fact on how well it is after it has been implemented. Getting people to use it, respond to it and update it is a big challenge. People generally don't take KM to be their day-to-day compulsory work. People complain about KM not giving them details when they require it. Proper training has to be imparted to effectively access the KM which is also difficult due to low number of experts in the field.

"What good is technology if it takes six seconds to send a message but six months to get someone to act on it?!"

8. *Access issues*: Establishing a KM network also involves defining roles of people who will be using and accessing the content. Without proper access checks, it becomes impossible to maintain the integrity of the data that is being stored in the database. Generally with huge companies and their databases spreading across nations, keeping a check on the knowledge flow becomes a big challenge.

9. *Funding and showing impact on business growth*: An initiative can get proper support from the management only when it shows that there is positive ROI from establishing such a unit. With KM this becomes the biggest challenge because sometimes results only start showing after a couple of years which is a significant delay in time if we consider the finance department where results are analysed every quarter. Generally KM measures have to be shown in a different

light, and other than financial aspects, a different scorecard has to be designed. For a moment here if we consider NPOs, for them it's a bigger challenge to show effectiveness of KM initiatives, and they also have a fund crunch which doesn't let them waste even a dime.

10. *Integrating KM into the culture*: The top management pushes for implementation of an efficient KM structure, but the idea doesn't gain popularity at the lower department levels. People don't realise the importance of KM. So, essentially, to get major benefits of KM, one needs to incorporate the KM habit as some organisations say. It has to develop in the subconscious of the employees to use KM and discuss new additions in KM. During my stint at Wipro Technologies, I have experienced the changes in my work design when KM builds into your daily routine. Writing same lines of code which have been formulated earlier is a waste of time. Using KM to get the same thing increases my productivity as well as the overall efficiency of the team.

4.3 Implementation Challenges of Knowledge Management

4.3.1 Cultural

- *Management support*: In the absence of management support, it becomes very difficult to carry forward an idea. Usually, KM initiatives are majorly started by a group of employees who form communities of practice, and they expand this idea of sharing knowledge for each other's benefits, and the management supports them and scales up the idea to form a knowledge management network.

- *Demonstrating business value*: Every activity which is carried on in the organisation should align itself to the business goals of the organisation and also reflect the value system of the company. In cases where the KM team falls apart from the culture and value system, KM finds it very difficult to sustain itself in the long run and can fall apart in case of emergencies.
- *Change management*: Usually with growing competition, the companies generally go through a revamp of their model of working and refocus on the business goals. These days it has become more of a dynamic process. Hence change is something which has become constant in any organisation. Hence it becomes very important for the KM team to continuously keep themselves updated and act according to the occurring changes.
- *Evolving technologies*: Technological aspect may very well be a part of the change management challenge. But it needs to be put separately as for implementation of a KM framework is highly dependent on the technology which is being chosen. With technological upgradation, keeping KM structure up to date is a major challenge not only because it involves a lot of efforts but also a lot of funds have to be poured in.
- *Security*: With lots of people getting access to the Internet and knowledge of the design being freely available on the Internet, saving data online has become a threat. There have been lots of instances where there have been security lapses, and the company has run into losses of millions of dollars. Even for smaller organisations which generally use open source software, it's a big problem because these software don't provide enough security.

4.3.2 Technology Infrastructure

- *Integrated databases*: Data for a company exists at different locations and in different formats. Integrating all these data and forming a robust structure to give professional benefits to the people at the organisation is a challenge. A lot of information lies without the proper context. Without proper backup the data becomes useless. There are lots of other technological road blocks like huge chunks data that can't be shifted without losing some data.
- *Interoperability*: A KM process should ensure that the person using it has the liberty to access in whichever way he/she likes. There has been a lot of debate about companies not letting the information which might be quite useful into the public domain. KM should be designed in such a way that the employee has the independence to access the information available on the Internet, i.e. the free domain, and also match that up to something that has happened at their own organisation.

4.3.3   Process and Architecture

- *Business process*: KM crusaders generally say that KM should be integrated with the business process of the company or the institute. But it is easier said than done. The architecture of the organisation and the structure influence the way in which KM is designed.
- *Impact analysis*: Aligning a monitoring and impact analysis module to KM has been a challenge throughout. Today's businesses run on targets and numbers. Places where numbers don't show don't really matter. Here it's difficult to measure the impact because it's based mostly on the subjectivity of the process. No one can be sure that the project succeeded because KM was used properly. Neither do people want to attribute their success to anyone else's work.
- *Integration into the budget*: Generally when the budget is processed at the beginning of every financial year, fund allocation is done according to the business goals of an organisation prioritising the departments. Till date most organisations have been unsuccessful in portraying KM's contribution to the organisation's growth. So lots of companies have a KM structure in place but are facing difficulty in scaling it up or restructuring it.

4.4   Adaptation of NPOs to KM

Before we get into details of challenges that NPOs face while implementing KM, let's first look on how they have managed to set up a structure in some organisation. The following section through some wide questions will analyse how the non-profit sector has benefitted out of it and how they are evolving in terms of knowledge management.

Can the non-profit sector benefit from knowledge management?

The non-profit sector has to undergo a lot of challenges which is unknown to the business world. There are a lot of stakeholders involved other than the shareholders in case of a NGO. The accountability of a non-profit doesn't depend on how much money they make so the simple ROI calculations to measure the benefits may not work out. Having said this, it doesn't mean that non-profits shouldn't have a KM system. There are international funding agencies like DFID who have successfully implemented them and are using it. Smaller organisations like New Concept Information Systems Private Limited also maintain a repository of documents of ideas which have been implemented, and in due course of time, they have found it to be very useful.

Is the non-profit sector consciously adopting knowledge management strategies?

The non-profit sector is undergoing an extensive renewal process. They are required to deliver tailored and high-quality services in order to overcome environmental complexity and scarcity of resources. Due to these changing expectations, they have started re-engineering their working culture and processes. They are putting in thoughts to adopt new ideas which have been successful in the business world. They have realised that intellectual property is something that they can build their credibility on.

What limits the non-profit sector from maximising benefit from knowledge management strategies?

The non-profit organisations often operate locally and are very specific to the mission. They also work with limited resources and financial constraints, and if they are working with some government agencies, they have to follow strict protocols. Hence decision-making and formulating plans are not easy for a NPO. Because of this and other reasons which are explained further in the article, these organisations have prioritised KM low on their lists.

Based on this, we will carry forward our discussion about KM challenges which are specific to the not for profit organisations.

4.5   Challenges Being Faced by the NPO Sector

4.5.1   Phases Crucial to Knowledge Management

Phases crucial to Knowledge Management:

Sourcing
- Rich and accurate foundation for knowledge formation
- Marking experts, customer feedback, published literature, staff contibution and ideation

Abstraction
- Principles, theories, and concepts to guide knowledge devvelopment

Conversion
- Abstract Concepts are convereted into applications and outcomes

Diffusion
- Shared understanding and adoption of the gained knowledge

Development
- Adaptable and flexible knowledge which accomodates changing contexts

4.5.2   Challenges for Each Phase

Challenges for each phase:

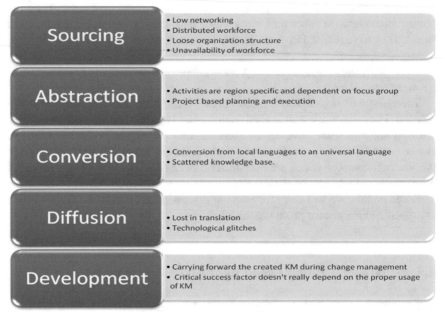

1. *Low networking*: The NPOs generally function in one specific region. Unless they are a big organisation with international funding, they tend to remain close to their work base. Thus the network they make among the other developmental workers is low. The information shared to them and by them remains low. This becomes a problem when an organisation is trying to assimilate data and put it together.

2. *Distributed workforce*: The workforce of NPOs remains scattered. So getting them together and forming COPs which form an essential part of KM is very difficult.

3. *Loose organisation structure*: The organisational structure is not very well defined, and majorly these have flat structures where a single person is in charge of multiple duties due to which it again becomes a problem to assign a KM job to some person. People generally tend to assume to be an overhead rather than some productive work.

4. *Unavailability of workforce*: Finding people to work for NGOs is still a major challenge in the mainstream work of NGOs. Proper training facilities are still unavailable. Most of the work is still driven by volunteers rather than professionals.

5. *Activities are region specific and dependent on focus group*: As the activities vary with projects and area of operation, developing a process is not feasible. Unless there is a process in place, assimilation work of KM is difficult to carry out.

6. *Project-based planning and execution*: All the planning is done at the grass-roots level, and documents are prepared, used and never submitted to a central system. As the system is quite decentralised and this has worked for the sector quite well, it is a challenge to change the system and come up with a new one.

7. *Conversion from local languages to a universal language*: Al transaction and conversation take place in the local language. Even the village level worker has limited knowledge of the English language which is the most accepted language for KM. So these people face difficulties in recording whatever they are doing for their projects.

8. *Scattered knowledge base*: The NPOs who are more than a 100 years old face a unique problem of a huge number of documents, but there is no way in which any particular information can be traced back. Multiple offices have kept same copies of the data and there is no data integrity in these organisations. This becomes a challenge when KM is in initial changes of its implementation.

9. *Lost in translation*: Most NPOs operate at areas where people speak a core dialect and all the documentation and information gathering in done in local languages. When this knowledge is transferred to English, a lot of meaningful data is lost during translation. Same sentence written in a different language might give an entirely different meaning when spoken in a local language.

10. *Technological glitches*: Establishing a KM structure is a costly affair. Smaller organisations who rely on grants and donations usually use open source software which limits the functionalities KM can offer to the employees. Besides the maintenance of high-end software, high-end software is also costly and requires expertise. These things can be quite challenging for a small NGO.

11. *Carrying forward the created KM during change management*: The development of organisations is changing very fast and so is the way they generally work. Including so many changes over a short period of time, it has left the NPOs in a very critical position at which they have to take a call on what to keep and what to let go.

12. *Critical success factor doesn't really depend on the proper usage of KM*: Each project in a NPO is very different from others. It varies according to the target group, the region and lot of other factors. So technically speaking, one can't use the process followed in another project and be assured that it will work. Measuring the success rate of a project from KM metrics becomes very difficult. The challenge again comes back to proper alignment of KM metrics to the CSF of a project. If all the metrics are being made, they have to be divided department wise which will have a lot of overheads.

4.6   Conclusion

With any new idea or technology, there have always been challenges and problems. The transition phase is very crucial as it goes through all the available test cases and rectifies itself and makes itself sturdy to face challenges for the future. All the organisational processes get aligned to the VMG of any organisation after going through a process of refinement. So there has to be lot of struggle to bring KM to mainstream business functions. Some organisations which are at a maturity stage have implemented KM principles and have failed. So other smaller organisations should not be disheartened if the ROI on the efforts isn't visible immediately.

4.7 Summary

Growing businesses bring forth new challenges, and this leads to innovation which can bring about change in the way we approach the problem in hand. One of such innovation is on its way and is taking the form of knowledge management. The revolution started among the top corporate of the world, but the revolution has percolated to smaller firms and even to the non-profit sector. A lot of companies are reaping benefits of efforts they had put into knowledge management. But the things have not been smooth throughout. KM initiatives have always faced a lot of challenges right from inception stage to implementation stage. Similarly when the idea got transitioned into non-profit sector, there were a lot of adaptation issues.

The concept of knowledge management has been well adapted by the business sector, but recently their importance has also moved into the non-profit world where the relevance is increasing as the credibility of these organisations is coming more under the scanner around the world. Moreover these organisations are necessarily data-centric or more appropriately knowledge-based organisation. They use a lot of historical data as well as similar cases from regions around the world. Due to expanding professionalism in the NPO sector, there has been a constant strive towards adopting new managerial paradigms. As a part of this, these institutions have come to realise that KM strategies might help them to attain a competitive edge over other organisations.

In this paper, issues which organisation face while working with KM have been discussed in general. Later it also deals in details about the challenges faced by the NGOs in specific. As the NGOs handle the problems differently, it is also important to understand their perspective. The importance and usage of KM can only grow like a forest fire if people understand the value of knowledge that gets accumulated every second. The human brain is sort of a KM system where the number of gyrations on the brain increases as we go on accumulating knowledge. So we need to follow the creator and develop proper storage for what we are learning today for the use of future generations.

Chapter 5
Designing Knowledge Management Strategy

5.1 Introduction

Knowledge management derives their power from the knowledge they use. When knowledge developers begin the building process, the first step is capturing tacit knowledge. The next step is to find a way to codify and organise knowledge into a form for others to use when needed. Getting the right knowledge to the right people at the right time is the whole idea behind knowledge codification (Fig. 5.1).

5.2 Knowledge Codification

As an emerging trend, knowledge management systems are slowly succeeding in standardising terminology and meaning for the common user. As a matter of procedure, after knowledge is captured, it is organised and codified in a manner amenable for transfer and effective use. Knowledge codification is organising and representing knowledge before it is accessed by authorised personnel. The organising part is usually in the form of a decision tree, a decision table or a frame. Codification must be in a form and a structure that will build the knowledge base. It must make it accessible, explicit and easy to access.

From a knowledge management view, codification is converting tacit knowledge to explicit knowledge in a usable form for organisational members. From an information system view, it is converting undocumented to documented information. Regardless of the view, codification is making corporate-specific knowledge (tacit and explicit) visible, accessible and usable for value-added decision-making.

© Springer International Publishing Switzerland 2016
S. Mohapatra et al., *Designing Knowledge Management-Enabled Business Strategies*, Management for Professionals, DOI 10.1007/978-3-319-33894-1_5

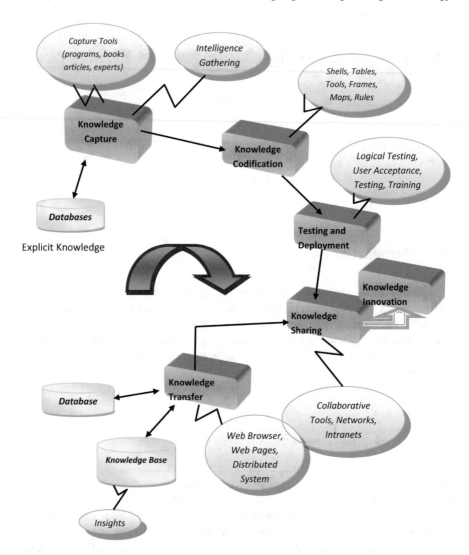

Fig. 5.1 Knowledge codification in KM cycle

5.2.1   Principles of Knowledge Codification

The following are the principles of knowledge codification:

1. *To decide what business goals the codified knowledge will serve*: It basically deals with defining the strategic intent of the organisation. The prime motive is to determine the business problem to be solved and align knowledge to be captured with business objectives.
2. *Identify existing knowledge necessary to achieve strategic intent*: It is usually very difficult to determine the knowledge requirements. A lot of subjective pro-

cesses raise political, cultural and strategic issues, thereby making the task furthermore challenging. Different experts have different perspectives about content
needs and sources of 'hard' and 'soft' information (ideas, gossip and opinion)
and thus contribute further to the complexity of the task.

3. *Evaluate existing knowledge for usefulness and the ability to be codified*:
   Evaluating the existing knowledge for usefulness of the information is very
   essential as only the knowledge that adds value to the repository should be
   retained. Only the useful knowledge, once identified, needs to be checked against
   its ability to be codified into useful information.
4. *Determine the appropriate medium for codification and distribution*: Media
   choice usually varies with the richness and complexity of the knowledge captured. It involves capturing information, filtering out the redundant information
   and further adding value to it through human input. While determining the
   appropriate medium, it is also necessary to organise and structure information in
   an accessible form. It becomes important to structure information appropriately
   as too much structure can hide information from employees whose mental modes
   don't match chosen structure. It is helpful to classify knowledge into the following categories for prompt structuring of information:

   • Process knowledge (best practices) that can increase efficiency.
   • Factual knowledge: easily documented but of little value unless synthesised
     and in context.
   • Catalogue knowledge: it shows where things are (e.g. yellow pages).
   • Cultural knowledge.

5.2.2   Codification Requirement

Organisations constantly adapt to changing situations—changes in market, changes
in customer tastes and preferences, turnover among company specialists and experts
and new developments in technology. In codifying knowledge, the resulting knowledge base serves in several important training and decision-making areas. Among
the important ones are the following:

(a) *Diagnosis*: A diagnostic knowledge management system is given identifiable
    information through the user's observation or experience. Built into the system's
    knowledge base is a list of all identifiable symptoms of specific causal factors.

    Example: *For a medical diagnostic system, the goal is to identify the patient's
    illness. It starts by assuming that the patient has a particular illness. It proceeds
    by reviewing the rules and their actions and asks the patient for additional
    information about unique symptoms in order to create a description of the
    illness. All previous responses by the patients are retained in working memory
    so that repetitive questioning of the patient is avoided.*
(b) *Instruction/training*: The basic concept of this type of codified knowledge base
    is to promote training of junior or entry-level personnel based on captured
    knowledge of senior employees. Most training knowledge management systems

have explanatory capabilities; they display the reasoning behind their solutions. Some systems allow students to pose hypothetical 'what if' scenarios, enabling them to learn by exploration.

(c) *Interpretation*: Interpretive codified knowledge systems compare aspects of an operation to preset standards. Typically, they use sensor data to infer the status of an ongoing process or to describe a given situation. Much of the reasoning is coupled with confidence factors.

Example: *If knowledge rule in an auto knowledge advisor reads, 'If water temperature is over 50% of max and oil pressure is less than 9, then there is 0.78 certainty that engine is low on oil.' In such a case, the system flashes a warning message, 'CHECK ENGINE OIL'.*

(d) *Planning/scheduling*: A planning knowledge management system maps out an entire course of action before any actual steps are taken. The system creates detailed lists of sequential tasks necessary to achieve two or more specific goals.

Example: *A system used by the US Air Force in the Afghanistan war. It makes 5-day projections of the enemy's military capability over a designated geographic area. Once a plan is completed, the system refines each step to arrive at the optimal schedule. The system rejects any out-of-line approach that is not within the prescribed constraints.*

(e) *Prediction*: Predictive knowledge management system infers the likely outcome of a given situation and flashes a proper warning or suggestions for corrective action. The system evaluates a set of facts or conditions and compares it by analogy to precedent states stored in memory. The knowledge base contains several patterns on which predictions are based.

Example: *Predictive systems are used in hurricane prevention. They are capable of estimating damage to real estate, and they can predict the nature of the hurricane by evaluating wind force, ocean and atmospheric conditions.*

It can hereby be summarised that justification for codifying knowledge is to address the areas where codified knowledge results in efficient, productive and value-added knowledge bases and knowledge applications. It means improving an organisation's competitive advantage by reducing its dependence on human experts, who are expensive, hard to come by, inconsistent and mortal. Codified knowledge and the resulting knowledge base mean allow organisational users to perform at a level closer to that of a human specialist in the domain. In the least, knowledge bases can be used as training grounds for junior employees to go through the thought process of the seasoned specialists and learn from it.

5.2.3 Points to Observe

Knowledge codification is neither easy nor straightforward. There are things for the knowledge developer to remember before beginning to codify. The following points should be considered:

- Recorded knowledge is difficult to access, either because it is fragmented or poorly organised.
- Diffusion of new knowledge is too slow.
- Knowledge is not shared, but hoarded. This has political implications, especially when knowledge cuts across departments.
- People are oblivious to who has what knowledge.

It is also crucial to understand that knowledge is often:

- Not present in the proper form (form)
- Not available when needed (time)
- Not present where the knowledge process is carried out (location)
- Not complete (content)

5.3 Modes of Knowledge Conversion

Before understanding knowledge codification, it is important to distinguish between knowledge capture and knowledge creation and how they are treated in the knowledge codification phase. The basic assumption in knowledge creation is that individuals create knowledge and that organisational knowledge creation is further amplification of the knowledge created by members of the organisation. Knowledge capture, in contrast, is viewed as a part of knowledge creation. In the process of knowledge capture, usefulness is often more important than originality and must be followed by adaptation and codification to the organisation's needs.

Tacit and explicit knowledge can be identified as two major types of knowledge. Both can be combined into four different modes of knowledge conversion. The four modes of knowledge conversion can be categorised as follows:

1. *Socialisation*: Tacit to tacit knowledge produces socialisation, where experience is what a knowledge developer looks for during the knowledge capture process. Observation and practice are two important tools that can be utilised effectively for capturing of knowledge.
2. *Externalisation*: From tacit to explicit knowledge is externalising. It can be described as explaining or clarifying tacit knowledge via analogies, models or metaphors. Explicit knowledge can then be stored in knowledge bases or repositories for day-to-day use.
3. *Internalisation*: From explicit to tacit knowledge is internalising or fitting explicit knowledge into tacit knowledge. Practicing again and again to imbibe certain techniques can be viewed as an example of internalisation.
4. *Combination*: From explicit to explicit knowledge is more or less combining, reorganising, categorising or sorting different bodies of explicit knowledge to lead to new knowledge. An example of this would be sorting an employee master file to see how many employees are older than 60 years of age, fall between the ages of 50 and 64 and so forth (Fig. 5.2).

Fig. 5.2 Modes of
knowledge conversion

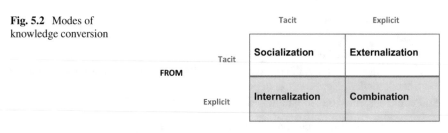

Knowledge capture is an integral part of knowledge creation, which requires group work, recognition of the importance of tacit knowledge and support for internalisation and socialisation. Knowledge creation also requires experience, which over time leads to expertise. Human experts become candidates for tapping their tacit knowledge via knowledge capture.

5.4   Points to Ponder Before Going for 'Codification'

It is a challenge to retain knowledge and its distinct properties. Any organisation, before attempting codification, must focus on the following aspects:

- What organisational goals will be served by the codified knowledge? Codifying puts knowledge in a usable form: a corporation should have a specific strategy for its usage. Knowledge, first-hand alone, is not enough. Different perspectives are required to serve the needs of different goals of a knowledge management project. At this instance, it is required to assess an organisation's needs and knowledge in order to develop a strategy for fulfilling its unfulfilled needs.
- What knowledge exists in organisation that can address these goals? It requires starting with a survey of numerous aspects of knowledge—availability, presence, use, transfer and so forth. At this stage, it is required to understand the knowledge and work practices of an organisation and subsequently using these insights to help transform the organisation into what it intends to become. It is important that such knowledge is identified and put into a usable form before sharing the same outside the organisation. The goal is to look for relevance of knowledge, not in completeness of the same.
- How useful can be existing knowledge for codification purpose? One way to make existing knowledge useful is by adopting knowledge maps (it involves diagramming the logic trail into a crucial major decision).
- How can one codify knowledge? It needs to be identified as to what medium or tool would be the most appropriate for codification? Mapping corporate knowledge is an essential part in the codification process. In order to accomplish the job, a knowledge developer must identify the knowledge type—schematic versus rich, tacit versus explicit, frame based or rule based and so forth. What can be achieved with knowledge depends upon its importance level; its nature, i.e. type; and how labour intensive it can be for proper evaluation.

5.5 Codifying Tacit Knowledge

Tacit knowledge is complex. Tacit knowledge is the type of knowledge that gets developed and internalised in the mind of a human expert over a long period of time. Due to this reason, it is not easy to codify tacit knowledge completely in a knowledge base or a knowledge repository. The distinctive style of the expert can barely be described or externalised in a set of rules as part of the codified knowledge. In short, it requires special knowledge-capturing skills to codify tacit knowledge efficiently.

It can be observed that certain knowledge can be more difficult to codify into rules and formulations in comparison to scientific or quantifiable knowledge. It would be futile and too laborious to process and codify an expert's year of intuitive experience. Hence, whenever a knowledge base project gets completed, it can only be referred to as being good enough for usage; it can neither be complete nor it can be final. Maturation with human knowledge over time continues to be the missing area in codified tacit knowledge.

Due to above-mentioned reasons, codifying tacit knowledge is accomplished by relying heavily on an expert with the knowledge and a knowledge developer to capture such knowledge. Dealing with experts is not easy, but once an expert is identified, a knowledge developer needs to understand and clearly access the expert's style of expression. The expert's ways of expressing his/her knowledge may vary. It can be explained by following a procedure, or understood as a story, or the salesperson type who sells his or her own way of solving a problem. The knowledge developer must adopt knowledge capture and codification methods in line with the expert's way of expressing his/her knowledge.

Many organisations lack in transparency in terms of sharing the knowledge across the entire hierarchy. Critical knowledge is often available, but it is difficult to be found and hence unexplored. Both knowledge identification and knowledge capture process improve knowledge transparency, which is a prerequisite for sharing of knowledge. In order for the knowledge to be identified, it needs to be codified. A codified form of knowledge needs to be created. In this respect, codification means representing knowledge in a useful and meaningful way so that it can be transferred and shared by any authorised user of an organisation.

5.6 Codification Tools and Procedures

Knowledge can be shared through personal communication and interaction. It occurs naturally all the time and is very effective; however, rarely the process is cost-effective. Knowledge codification is the next stage of leveraging knowledge. Through conversion of knowledge in a tangible explicit form such as documents, it can be communicated more widely without incurring much cost.

There are numerous costs and challenges associated with knowledge codification. The first issue is that of quality, which encompasses:

1. Accuracy
2. Readability/understandability
3. Accessibility
4. Currency
5. Authority/credibility

Knowledge codification serves the pivotal role of allowing what is collectively known to be shared and useful. People always used some type of knowledge codification during their everyday activities to make communication and discussions more effective. Work or business jargon, email and computer programmer's technical language are only some examples. However, it is impossible to codify in a document or a database the knowledge, skills, expertise, understanding and passion of an employee. In this case, the best solution is to provide a link to the sources of knowledge using knowledge maps, company yellow pages or a company guide. If people interact to share their knowledge within a community of practice, then that practice becomes more effective. If knowledge is codified in a material way (i.e. it is rendered explicit), then it can be shared more widely in terms of both audience and time duration. Knowledge must be codified in order to be understood, maintained and improved upon as part of corporate memory.

The codification of explicit knowledge can be achieved through a variety of techniques such as:

• Cognitive mapping
• Decision trees
• Knowledge taxonomies
• Task analysis

5.7   Cognitive Maps

Once expertise, experience and know-how have been rendered explicit, the resulting content can be represented as a cognitive map. A cognitive map is a representation of the 'mental model' of a person's knowledge and provides a good form of codified knowledge. In the map, the nodes represent the key concepts, while the links between them show the interrelations between concepts. Thus, cognitive mapping is based on concept mapping and allows experts to construct knowledge models. They could show multiple perspectives or views on the content.

Cognitive mapping is based on concept mapping through which experts can directly construct knowledge models. Concept maps represent concepts and relations in a two-dimensional graphical form, with nodes representing key concepts connected by links representing prepositions. These are quite similar to semantic networks used by diverse disciplines such as linguistics, education and knowledge-

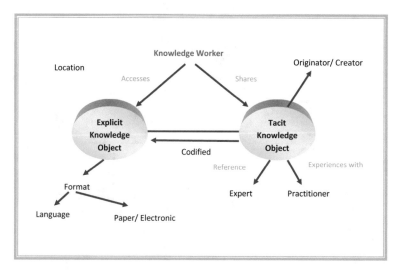

Fig. 5.3   Example of concept map

based systems. The goal of such systems is to better organise explicit knowledge and to store it in corporate memory for long-term retention (Fig. 5.3).

5.8   Decision Trees

Decision trees are another widely used method to codify explicit knowledge. This representation is both compact and efficient. A decision tree is typically in the form of a flowchart, with alternate paths indicating the impact of different decisions being made at that juncture point. A decision tree can represent many 'rules', and when you execute the logic by following a certain path, you are effectively bypassing rules that are not relevant to the case in hand. The graphical nature of decision trees makes them easy to understand, and they are obviously very well suited for the coding of process knowledge (Fig. 5.4).

5.9   Knowledge Taxonomies

Concepts can be viewed as the building blocks of knowledge and expertise. Once key concepts have been identified and captured, they can be arranged in a hierarchy that is often referred to as structural knowledge taxonomy. Knowledge taxonomies allow knowledge to be graphically represented in such a way that it reflects the organisation of concepts within a particular field of expertise or for the organisation in its entirety.

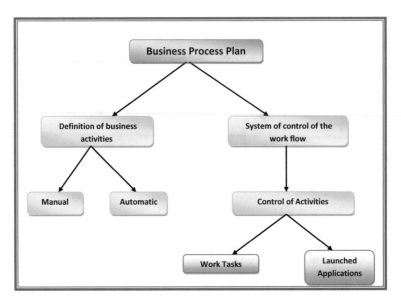

Fig. 5.4 Example of decision tree

Taxonomies are basic classification systems that enable us to describe concepts and their dependencies—typically represented in a hierarchical fashion. The higher up the concept is placed, the more general or generic the concept is. The lower the concept is placed, the more specific an instance it is of the higher-level categories. This approach allows lower or more specific concepts in the taxonomy to directly incorporate the attributes of the higher level or the parent concepts (Fig. 5.5).

The different taxonomic approaches to the codification of explicit knowledge are summarised in the table below (Table 5.1):

5.10   Strategic Implications of Knowledge Capture and Codification

Knowledge capture and codification are particularly critical when an issue of knowledge continuity arises. KM is concerned with capturing and then sharing know-how valuable to colleagues performing similar jobs throughout a company. Knowledge continuity management stresses on passing critical knowledge from existing employees to their replacements. Most of the available literature focuses on transfer of departing individual knowledge to successors of departing individuals; the problem however is not so localised and needs a broader outlook. Knowledge continuity should not only focus on the specific knowledge to be transferred between individuals but in addition should address strategic concerns at group and organisational levels. The organisation needs to be aware of its critical knowledge assets; these are captured and codified in

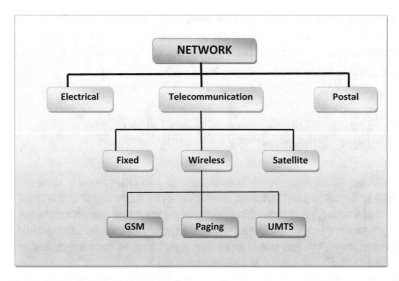

Fig. 5.5 Example of knowledge taxonomy

Table 5.1 Major taxonomic approaches to knowledge codification

Taxonomic approach	Key features
Cognitive or concept map	• Key content represented as a node in a graph, and the relationships between these key concepts are explicitly defined
	• Can show multiple perspectives or views on the same content
	• Fairly easy to produce and intuitively simple to understand but difficult to use for knowledge related to procedures
Decision tree	• Hierarchical or flowchart type of representation of a decision process
	• Very well suited to procedural knowledge—less able to capture conceptual interrelationships
	• Easy to produce and easy to understand
Manual knowledge taxonomy	• Object-oriented approach that allows lower or more specific knowledge to automatically incorporate all attributes of higher level or parent content they are related to
	• Very flexible—can be viewed as a concept map or as a hierarchy
	• More complex; will therefore require more time to develop as it must reflect user consensus
Automated knowledge taxonomy	• A number of tools are now commercially available for taxonomy construction.
	• Most are based on statistical techniques such as cluster analysis to determine which types of content are more similar to each other and can constitute subgroups or thematic sets
	• Good solution if there is a large amount of legacy content to sort through
	• More expensive and still not completely accurate—will need to validate and refine for maximum usefulness

the knowledge map or taxonomy. Organisations also need to take into account the impact of a separation/departure on the communities to which they belong, whether the departure is due to retirement or other reasons. Such departures may literally leave a serious gap in the fabric of the community network.

5.11 Practical Implications of Knowledge Capture and Codification

The benefits of capturing tacit knowledge and codifying explicit knowledge are obvious to organisations; however, they seem to be fairly vague at the level of the individual knowledge worker. Knowledge is an asset that cannot be owned but is merely borrowed or rented. Some knowledge remains within the organisation when employees leave, but this needs to be the 'right' kind of knowledge, and workers will need to be able to access and make use of it.

Recommendations for promoting knowledge capture and codification are as follows:

- *Acknowledging knowledge contributors*: Turning tacit knowledge into explicit knowledge is difficult for many users and often meets with resistance, despite the obvious benefits. It is important to acknowledge workers who not only create original content but also help improve the content over time by adding context from customer interaction. KM software should offer reports to identify those who are contributing or help to tap the tacit knowledge by building profiles of experts based on their contributors.
- *Unlearning*: The role of unlearning or reframing is of utmost importance. The organisational knowledge base should not be viewed as unlimited storage space to be filled. Although there may not be any technological constraints, certainly some conceptual constraints need to be taken into consideration. Unlearning involves disposing of old frameworks and breaking away from the status quo—a form of double-loop learning.
- *Preventing loss of knowledge during transfer*: Conversion of tacit to explicit knowledge must be accomplished without any significant loss of knowledge during the process. The advantages of communicability do not always outweigh the disadvantages of 'knowledge leaking'. It is just as valuable to maintain a link to knowers—individuals within the organisation who are adept at making use of complex knowledge. The goal is to carry out the 'right' amount of knowledge acquisition and codification.
- *Remember the paradox of knowledge value*: It needs to be understood that the more tacit knowledge is, the more value it holds. Tacit knowledge is generally of greater value and of greater competitive advantage to a firm. It may be in the firm's interest to maintain that content at a certain minimal level of tacitness so that it is not easily acquired or imitated by others.

5.12   Case: Knowledge Management 'Dell (Formerly Perot Systems)'

The case elaborated below covers the objectives and outcomes of Dell Services KM initiative in the area of 'project management'. It details the views of Mrs. C. S. Shobha, Head—KM Quality and Training, at Dell Services.

The following constitute Dell Services KM objectives:

– Elevating knowledge work towards a key focus direction
– Maximising the importance of collective learning in project teams
– Integrating KM into daily workflow
– Delivering innovative solutions to customers in a sustainable manner and making the organisation an attractive choice for knowledge workers

Initially, knowledge audit at Dell services was driven by the '8 Cs' framework: connectivity, content, community, capacity, culture, cooperation, commerce and capital.

As per Mrs. Shobha: 'KM has helped in project management performance. DELL created part-time roles for each project called Knowledge Champions, chosen by project managers. Knowledge Champions received induction kits about expertise mapping, facilitating knowledge sharing sessions, preparing case studies, and focusing on reusability and innovation'.

KM in projects helped build testimonials/credentials for customers (new services as well as upselling/cross-selling). The numerous dimensions of knowledge management in project ecosystems include: domain knowledge, customer knowledge, methodology, technology, applications and project management. Different flavours of knowledge management are required as per changing project environments. In Dell, knowledge management has proven itself useful at all project phases: initiation, planning, execution, feasibility analysis and closure—including best practices, expertise maps and reusable artefacts.

Knowledge management has assisted in creating a huge perception shift in the way project and service managers were regarded. Now 'delivery heads' are referred to as 'knowledge leaders', 'project leads' as 'knowledge champions' and 'team members' as 'knowledge workers'. Dell's knowledge portal is called 'K-Edge', and there are 30 existing community of practices, some of which mandate involving customers in cocreation of knowledge assets and the knowledge management plan. The organisation also has an 'ideas management 'tool' which is utilised for rating and validating innovations. Presently, knowledge management metrics and assessment include percentages of employees involved in various knowledge activities.

Leadership needs to play different roles in different stages of knowledge maturity. In the beginning of knowledge initiatives, strong initial support of organisational leaders is sought for the launch. The tricky part is explaining the 'return on investment' aspect of implementing such initiatives in the organisation.

At Dell, knowledge management has become a part of project compliance. The company organises 'knowledge champion conclaves' every quarter in a year.

Through knowledge management initiatives, many evolved into better project managers. Now at an increasing rate, more and more employees want to become 'knowledge champions'.

Knowledge management initiatives at Dell have transformed the perception of their employees towards knowledge. Employees are better oriented at capturing and sharing knowledge within the organisation. Dell has successfully imbibed the culture of 'knowledge management' in its practices.

5.13   What Is the Importance of KM Strategy

Knowledge is a set of information which provides ability to understand, differentiate and view situations from different perspectives and further enables to anticipate its implications, judge effects and suggest ways to handle the situations. Knowledge has six characteristics, viz. expandable, compressible, transportable, sharable, diffusive and substitutable. It is dynamic and built out of perception, skills and training. It is termed as actionable information that has economic value. Knowledge has become essential for business and its management. It helps in building innovative and creative strategies for the organisation providing it competitive advantage over others. Thus it is considered as an asset.

5.14   What Is KM Strategy?

Knowledge management is a system's approach to identify, validate, capture and process knowledge and then organise the knowledge elements into knowledge assets for business function operations and decision-making. Knowledge management addresses several problems of the organisation like ageing workforce, disturbance in human resource balances due to mergers and acquisitions, unplanned need of expertise, etc. It can be broadly defined as the systematic and explicit management of knowledge-related activities, practices, programme and policies within the organisation to produce competitive quality of knowledge and its efficient and effective application in business management.

5.15   Importance of Knowledge Management Strategy

Importance of knowledge management has been felt since the evolution of mankind and it's ever increasing. It has become quite indispensable in today's fast-paced world where globalisation and liberalisation are the buzzwords. The systematic study of knowledge management, as a management and scientific discipline, began in 1994, with the annual report of leading Swedish financial services firm, Skandia, who made an effort to quantify the value of intellectual capital in his company. With this attempt, it was realised that the intellectual capital of an organisation is equally important to financial capital or other assets like labour or land as it imparts sustainability to the organisation. KM is also important as in this century there has been a revolution in usage of technology and the way business is done. Also the style, culture and structure of business management have changed giving rise to new business models. This has led to highly competitive business world where sustainable competitive advantages are necessary to survive. Such advantages are only possible with effective management of knowledge.

5.16   Vision, Mission, Goal and Its Importance for Knowledge Management Strategy

Vision, mission and goal are very important for an organisation, since they are primarily responsible to strike an alignment between the employees and motivate them to work in sync with each other and the organisation to achieve the desired outcome.

A *vision statement* outlines what the organisation wants to be or how it wants the world in which it operates to be. It concentrates on the future. It is a source of inspiration. It provides clear decision-making criteria. A *mission statement* tells you the fundamental purpose of the organisation. It defines the customer and the critical processes. It informs you of the desired level of performance.

The advantage of having a stated vision, mission and goal is that it creates value for the employees, managers, etc. in the organisation. They create a sense of direction and opportunity and thus are very essential for the strategy-making process. Their contribution in the ongoing process of formulating, implementing and controlling broad plans to guide the organisation in achieving the strategic goods given its internal and external environment can hardly be ignored.

Vision/mission part (concluding): The importance of vision/mission can hardly be ignored as far as devising a KM system is concerned, since KM system needs to be perfectly aligned with the organisation's vision and mission statement. Only an alignment between the business strategy and the KM strategy can help an organisation achieve its objectives. The association of business strategy with the KM design, along with the implementation strategy, will ensure that the objectives of the implementation are met and ensure the successful completion of the project while increasing the capital of the organisation, both intellectual and financial.

5.17   Knowledge Management: A Strategic Initiative?

The kind of change the knowledge management initiative intends to bring is not a small one, nor is it a kind of project that can be taken, dealt with and then kept aside. It is something that is done for maintaining the knowledge that exists in a particular organisation and that needs to be constantly viewed and upgraded since it works directly in the interest of the organisation and its employees and helps in achieving the goals of the organisation. Since knowledge is an asset to the organisation, the creation process focuses on discovery of that knowledge which is critical to the organisation's success and performance. If the organisation's aim is to become lean, flexible and highly responsive to customer needs, then KM initiative is to be treated as 'strategic'. Similarly, if the organisation's aim is to be competitive and be ahead of the competition by leveraging on knowledge and knowledge assets, then KM initiative is 'strategic'. KM is a strategic endeavour and hence it is very important to link knowledge and KM to business strategy. However, such a strategy can be more right if the customer's requirement knowledge is captured through KMS. Further, the KMS which can generate such knowledge may demand technology infrastructure which will identify and articulate customer knowledge to build sustainable competitive advantage. The process is given below:

1. Begin with key knowledge area: The question 'where to begin?' would arise when KM initiative decision is taken. There are seven knowledge areas, like customer knowledge, product knowledge, relationship knowledge, process knowledge and so on. Which one to choose? The best way to handle this issue is to choose the area which is most useful and beneficial and has all the ingredients of successful launch.
2. Focus on explicit knowledge: We have two knowledge types, namely, explicit and tacit knowledge. The character of explicit knowledge, however, is more suitable to focus and start a KM initiative. Explicit knowledge compared to tacit is clear in definition, has a specification, can be coded and stored and is easy to disseminate. It can also be handled by technology at a relative ease and hence is the choice starting the initiative.
3. Set realistic expectations on KM: The expectations must be measured ones and duly consider all probable limitations and issues in KM implementation. The expectations should be set keeping in mind the customers and the end users' expectations who are not only users but also the stakeholders in the KM initiative. It is better to start with lower but realistic expectations and then raise them as KM becomes a reality.
4. Integrate KM into existing systems: KM systems should be an integral part of the larger information system portfolio for better effectiveness. It should be an integral part of CRM, email, remote diagnostics systems, DSSs, helpdesks, etc.
5. Educate users; internal and external: All the users must be aware of the KM system usage. They should be exposed to the total scope of KM systems and its internal which would help them to contribute to its sustained growth.

5.18   The KM Team

For an effective KM team, the sills, attributes and backgrounds that need to be possessed by the members must be clearly defined. The KM skills required can be grouped under seven categories, viz. retrieving information, evaluating information, organising information, analysing information, presenting information, securing information and collaborating around information. Similarly, the knowledge management roles must be defined in an organisation. However, the challenge lies in defining the key deliverables of those roles and specifying the skills and expertise expected in people who can perform those roles.

How to build a successful KM team:

- *Strong support*: A KM team needs strong senior management support as both the KM programme and the team are investments. KM programme needs to be started only when the organisation actually feels its need.
- *Innovation and exchange of ideas*: KM best practices are rarely repeated. Km team members need to innovate as well as exchange ideas with those who are already having experience in the field. Whenever a KM programme is initiated in an organisation, it shall have the flexibility to adapt to ongoing changes. The organisation shall also be open enough to start afresh if the initiative becomes unsuccessful.
- *Managing KM operations*: Each initiative must be governed and managed efficiently and it is just as crucial to start planning ahead. Without proper governance, the KM team will be consumed by the post-implementation duties.
- *Communicate successes*: Establishing baseline for the state before the implementation and measuring the effectiveness of the KM programme after the implementation can help motivate the team. The success of the KM team must be acknowledged which will help to build credibility for the team.
- *One of us*: To keep KM team members relevant to their colleagues, the KM team members must be up-to-date with what people in the organisation are doing, so that the KM team remains relevant to their colleagues.
- *The desired KM team*: Right kind of attributes and attitudes of the KM team members plays a much more important role than the KM skills or experience possessed by them. The KM sector is constantly changing. For a KM team to stay competitive in this dynamic scenario, it shall constantly keep up with evolving needs of the field, tap in new opportunities, unlearn old practices and gather new ones, exchange and validate ideas, form new relationships and demonstrate willingness to learn new things as an everyday affair. Skills can be picked up, experience can be accumulated, relationships can be established and artefacts can be collected; but attributes and attitudes are ingrained. In short, a KM team member must possess the drive for self-improvement and solve problems (Fig. 5.6).

5.19 Communities of Practice

Once knowledge has been captured and codified, it needs to be shared and disseminated throughout the organisation. While the tacit knowledge is hard to access, explicit knowledge which is usable and comprehensive is not easy to find. Thus, there are chances of the wastage of time and effort in reworking the information,

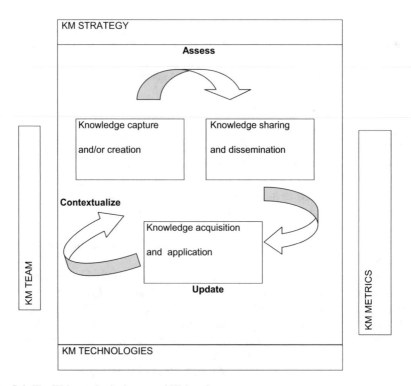

Fig. 5.6 The KM team in the integrated KM cycle

and even worse are the consequences of making decisions based on inadequate information. It has been found that other people are preferred source of information, as the information is fast and trusted. So interacting with other people not only provides a valuable activity but also enables a tacit-tacit knowledge transfer. Thus, it can be said that learning is a social phenomenon. Nowadays, technology is offering new media to employees with common interests or problems for sharing the knowledge, e.g. email groups, discussion groups, etc.

A community of practice refers to 'a group of people having common identity and professional interests and that undertake to share, participate and establish a fellowship'. The term community implies that these groups are not within the boundaries of geography and business unit. They are tied together with common tasks or interests. The word practice refers to 'knowledge in action' which highlights 'how individuals actually perform their jobs'.

A community of practice may be top-down as well as bottom-up. A COP has a mission statement and code of ethics wherein it produces results that add value to the organisation and bring improvement to the common professional theme of the members. However, the way in which they are formed when not top-down is quite unlike a professional organisation, as they self-organise and emerge on their own in a bottom-up manner. In order for effective knowledge sharing in a COP, a number of key roles need to be in place such as knowledge sponsor, champion, facilitator, practice leader, etc.

The communities help drive strategy for the organisation and may start new lines of business. They solve problems quickly and enable the transfer of best practices while developing the professional skills. They may play an important role in recruiting and retaining talent that may help expand the business. Apart from that, they facilitate the cross-fertilisation of ideas and building up of organisational memory. The communities also help build the social capital of the organisation. The communities of practice are increasingly becoming important as knowledge becomes a strategic intellectual asset in today's world.

5.20 Knowledge Management Process

Knowledge management involves identification and organisation of knowledge and knowledge assets for sharing with the knowledge workers to maintain competitive necessities. It is a critical link between knowledge assets and superior business performance. It is reflected in business strategy, policies and practices wherever it is considered as an important function. It is a way to improve the organisational results and learning by imparting specific processes, expertise and intellectual capital. It also makes possible the transfer and reuse of knowledge assets across the organisation. The processes mentioned from 'creation to use and exploitation' are very important to ensure that the knowledge shall be diligently added to the knowledge database. So the processes have a unique and important role to play in the knowledge management cycle. These processes are necessary to ensure quality and vitality of knowledge. *The knowledge management process involves the following steps*:

1. *Identification*: Knowledge needs to be relevant for the organisation and useful for its better management.
2. *Creation*: Identification, formulations and recognition of the source of knowledge along with a vitality check for the knowledge are performed over here. It is a cyclic process and contributes to the development of the knowledge base of the organisation. The forms of the organisational knowledge created are competitor knowledge, customer knowledge, supplier knowledge, product knowledge, technology knowledge, process knowledge, etc. The consistent component of this type of knowledge is the organisational context, and the ways in which they are created are referred to as knowledge creation cycle.
3. *Aggregation*: Knowledge is assembled and clustered into a unified set that would be easy to code, share, store and use. This is required to manage explosive amounts of information effectively. Although indexing and linking documents and other information sources is an important step, capturing the knowledge contained within these sources is crucial for building and using organisational information repositories.
4. *Organisation*: It includes the structuring of knowledge into organised formats so that it is easily accessible to users. It refers to the description of documents, their contents, features and purposes and the organisation of these descriptions so as to make these documents and their parts accessible to persons seeking them or the content that they have.

5. *Store*: The knowledge generated needs to be stored in easily accessible structure and format.
6. *Share*: It is a process to disseminate knowledge to users as well as ensure its security. Vast amount of information generated through routine process are lost on account of the inability of current people, processes and systems to manage, interpret and act on it. Unless knowledge is effectively shared and acted upon, it does not possess intrinsic value for an organisation or individual. This calls for increased need to access and share extra-disciplinary knowledge and to engage in meaningful transdisciplinary activities. This requires making knowledge explicit and facilitating distribution, analysis and gathering of new knowledge.
7. *Use*: This process keeps track and watch on knowledge usage. The knowledge must not be misused and must be utilised for the growth of the organisation and learning of the individuals.
8. *Validity of knowledge*: Validation of knowledge is required at specific intervals to maintain its relevance in the changing contexts as well as to enrich it. The processes used to create, communicate and apply knowledge result in expansion of the knowledge base that continues in a cyclic manner. However, the validity process ensures that there is no information overload and the knowledge that is important to the organisation is maintained (Fig. 5.7).

5.21   Knowledge Management Architecture

With knowledge taking on increased importance, it makes sense that there is an opportunity to create competitive advantage by effectively managing its storage and use. Effective knowledge management architecture creates competitive advantage by bringing appropriate knowledge to the point of action during the moment of need. Employee turnover is also reduced because a large portion of the knowledge and expertise acquired by the employee is captured in the knowledge base.

5.22   Principles of Knowledge Management and Design of Architecture

At this early stage of KM in business, the most appropriate form of dialogue is high-level principles instead of detailed tactics. Detailed approaches and plans based upon the principles of KM can be designed by the organisation.

The following are the key principles of KM that are important considerations while designing the architecture:

- *Scalability*: Architecture must be flexible enough to grow for accommodating increased number of users and processing components.

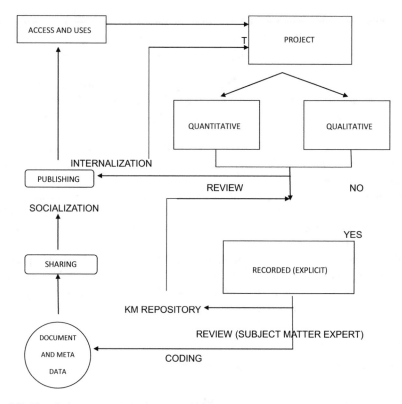

Fig. 5.7 Knowledge management process architecture

- *Security*: The components of the architecture shall have a degree of independence which refers to the ability of the components to make some decisions and send alerts. It shall automatically check the security status of the user.
- *System upgradation*: The architecture must allow restructuring of the existing system.
- *Knowledge exchange*: There must be steps to encourage knowledge exchange, develop a global knowledge organisation and build effective customer relationship.
- *Sizing and performance*: A reasonable understanding of volumes of data and its expected usage is a must to create adequate technology infrastructure.
- *Design*: Appropriate technology and usable and reliable system shall be used to ensure the ease in usage by the end user.
- *Investments*: An effective knowledge management system can be built by investment of other assets.
- *Hybrid solutions*: Mixture of skills is required to perform different tasks at organisations, and thus humans as well as technology find utility in different areas. Thus, a hybrid KM environment is needed to employ both man and machine in a complementary fashion.

- *Management and leadership*: There must be clarity in roles and responsibilities regarding the knowledge management in an organisation.
- *Market knowledge*: Mapping organisational knowledge can provide better access to the information.
- *Knowledge sharing and reuse*: For building a knowledge-sharing culture in the organisation, it is not only important to codify and enter it into the system, but it's equally important to motivate people to share their knowledge with others as well as benefit from others' knowledge.
- *Improved work progress*: For improving the KM process, it's important to make improvements in the key business processes like market research, product design and development, etc.
- *Organisational involvement*: Attention and engagement of the people of the organisation is quite indispensable for successful KM which can be achieved by summarising knowledge and reporting it to others, engaging in games based on knowledge usage, etc.
- *Cyclic process*: The process of managing knowledge is never ending as the knowledge is continuously changing with new technologies and management approaches. The mapping of a particular knowledge environment must be quick and only as extensive as necessary for productive usage.

5.23   Various Levels of KM Architecture

The guiding architecture of an organisation to be successful must be based on the strategic vision of the organisation. For a sustainable competitive advantage of the organisation, synergy among the different capabilities of the organisation at different levels is a must.

The architecture includes the following:

1. *Business architecture*: The business strategy of the firm is defined on the basis of the strategic vision. The business architecture takes into consideration business strategy of the firm, long-term goals and objectives, technological environment and external environment. It also takes care of interests of the stakeholders like government, regulatory authorities, customers and employees.
2. *Data architecture*: The data management strategy is based on the information and business strategy. The firm shall be clear about how the data will serve its business and information needs. The need for collection, usage and transfer within and outside must be defined. This level aligns various data-related aspects with the business applications, protocols and various hardware and software services. The following aspects shall be considered from a business perspective:

   - Database to facilitate structured storage of environment data, coupled with intrinsic operational data as well as data related to the organisation's external environment
   - Data dictionaries

- Database usage for intelligence mining
- Data mining to gather data on customers and competitors
- Data protocols
- Distributed database to provide a common view of data across the firm
- Data integrity and security
- Data warehousing
- Data modelling tools
- Development tools like Lotus Notes

3. *User interface layer*: This is the layer within the KM system architecture that has direct interface with the user. The success of the KM system depends upon the effectiveness of this layer. It supports platform independence, leverages organisational intranet and optimises video and audio streaming. There are certain basic needs that must be satisfied by the collaborative interface for effective collaboration across the enterprise, smooth sharing as well as presentation of knowledge from the organisational database. The major considerations are as follows:

- The collaborative platform deployed must be able to operate in a portable across all these platforms. The browser is the most suitable universal client through which end users can access the repositories of information without switching platforms or operating environments.
- Consistent and easy to use.
- Collaborative platform must have the scope of scaling up without degradation in performance as the number of the users increases.
- Knowledge sharing platform must integrate well with existing systems and applications.
- There must be reasonable scope of customisation and flexibility so as the end users can filter out irrelevant content and avoid the information overload.

5.24   KM Cycle

Knowledge management cycle plays a very important role to strategise the whole procedure of knowledge management. It deals with the process of capturing and organising knowledge for its transfer to users who can share and use it effectively. It also is a measurement for the knowledge performance and a way of evaluation for its continuity in the organisation. The cycle consists of four components, namely, conceptualise, plan, contextualise and improve (Fig. 5.8).

1. *Conceptualise*: Conceptualisation is the first step of the KM cycle wherein all the processes of knowledge management need to be found and the whole framework needs to be defined. Conceptualisation should be done in a way that it leads to the alignment of the longer and shorter business goals with the proposed KM strategy. Needless to say, KM strategy would not only influence the business strategy but also the future business planning procedures. Hence, the conceptualisation stage

Fig. 5.8 KM cycle

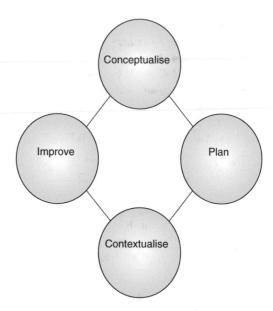

would involve a reliable method of identifying the existing knowledge and capabilities within the organisation that needs to be preserved and maintained at all levels for the benefit of the employees and organisation as a whole.

By the end of this stage, the KM framework should become visible to all key stakeholders. This would propose a vision of what and how this approach will hold mutual benefit. It is suggested generally that this vision should be formed before one actually ventures out in the planning stage as in the absence of a preset vision, the planning and implementation alone would not be able to actualise the whole initiative.

2. *Plan*: Planning stage involves in the alignment of KM with strategy and subsequently with the organisation's capabilities and resources. The transformation of the systems-level KM framework vision into a proper KM strategy takes place in this stage only. The strategy has to encompass all aspects of knowledge from capture, transfer and through to generation.

At this stage, the organisation also can start collecting knowledge and information about key competitors and markets and thus develop contingencies to hedge themselves from future adverse scenarios. Such a planning would help the organisation to model potential solutions and counteractive methods to deal with future situations, besides enabling them to prioritise their actions.

3. *Contextualise*: The next stage is the contextualisation stage which is all about situated application and where the activation of user access and enabling infrastructure takes place. For the user, this means the knowledge architecture exists to make knowledge easily accessible, highly personalised, exact, relevant and apt, available in communities and collaborative systems, have easy retrieval processes, highly controllable, etc. These activities intend to increase and improve the flow of knowledge to different individuals and teams involved in the organ-

isation. This has to occur to meet the knowledge requirements set by the organisation, customer demands and performance requirements of the employees.
4. *Improvement*: The last and concluding stage of the KM cycle is the improvement stage which provides the feedback provision and also marks an end of the KM execution cycles. In this stage, the final assimilation and merger of all the discussions and reports on outcomes from employees, teams, business partners, company leaders, etc. is done. This stage basically seals the fact that the proposed KM cycle is competent enough to achieve the targets and measures that are set in the KM strategy, thus leading to improvement.

5.25 KM Implementation Roadmap

Organisations that are looking ahead to become knowledge enabled start off by trying to become a learning organisation. The immediate steps that organisations take to this effect are to buy and attempt to deploy state-of-the-art KM tools and/or learning management systems (LMS). However, this results in cropping up of many problems like deployment, usage and adaptability to a learning culture. Therefore, to avoid such problems, an organisation can adopt a four-phased implementation strategy to design, develop and deploy an effective KM system. They are:

Awakening phase: In this phase, a review of organisational vision, mission and goal is done to ensure that the KM is absolutely aligned with them so that it can help the organisation achieve its objectives. Along with the alignment of business strategy with the KM strategy, infrastructural evaluation is also done in this stage that facilitates the understanding of various components that constitute the KM framework.

Actionable phase: The ideas generated in the awakening phase are given a concrete shape in the actionable phase. The second phase of implementation involves analysis, design and development of the KM system. There are five steps that constitute this phase, namely, knowledge audit, team formulation, architectural design, organisational K-print and system development.

Deployment phase: This phase involves the process of deploying the KM system that is designed in the preceding phases. This phase involves two stages: the first stage involves the deployment of the system with a result-driven incremental technique, and the next stage involves bringing about the cultural change and revised reward structure to foster knowledge sharing and ensure commitment of the employees and the desired ROI.

Maintenance and measurement phase: This is the last phase of the implementation process where the crucial process of measuring the returns of the effort as well as the process of measuring the business impact of KM needs to be taken up using a set of appropriate metrics. This includes metrics to facilitate the calculation of ROI for the KM investments as well as the usage of benchmarking techniques for deriving comparative metrics for the organisation. The measurement system should however account for both financial and competitive impacts of KM on the organisation's business.

5.26   KM, Organisational Culture and Change Management

5.26.1   KM Culture Creation

It has become something of a truism to describe knowledge management as primarily a cultural construct. To devise a good KM strategy, an organisation should try to cultivate and foster a good culture among its various departments. Culture may be characterised in multiple ways. However, one of the fundamental prerequisites of culture that acts as an obstacle to KM is the issue of trust. Knowledge sharing is enhanced when the employees feel that they are respected individuals and would be treated in a professional manner and that other members of the group can be trusted.

Further, proper implementation of KM also involves a change in the existing culture within the organisation, if not a complete transformation. Corporate culture is a key component of ensuring that critical knowledge and information flow within an organisation. The strength and commitment of a corporate culture will almost always be more important than the communication technologies that are implemented to promote knowledge sharing. However, culture creation is a very difficult task to do as cultures are deep-rooted, pervasive and complex. To bring about a change in the existing culture, the classic method of three steps is being used, namely, unfreezing, cognitive restructuring and refreezing. The key issue for leaders here is to become sufficiently marginal in their own culture to recognise, appreciate and assimilate the new culture and thus foster some new ways of thinking, thus bringing about a change in the organisation.

5.26.2   Rewards and Recognition Scheme

For an effective KM system, it is very crucial for an organisation to have an adequate and continuous knowledge sharing among its employees. The knowledge sharing across cultures, departments, etc. would not only lead to the creation of an environment of bonhomie and knowledge but also develop a sense among the workers that they are important assets to their organisations. However, the employees should have a reason to share their knowledge with others. Incentives, rewards and recognition thus remain one of the more important challenges facing KM today. Incentives are, however, tricky enough as its effect may be different on different people. For example, in an organisation, a person gets the award of 'the best employee of the year' for sharing and posting his knowledge on e-group to motivate other employees to take a similar course of action in the future. However, some of the employees may feel that as highly educated professionals, they should not be reduced to something that reminds them of a plaque used by fast-food companies to motivate their staffs. Therefore, an organisation needs to cautiously adopt the methods of rewards and recognitions which may be remunerative or financial or even moral in nature.

5.26.3   *Importance of Change Management in KM*

Knowledge management is rooted in the need for change—true KM requires a shift in culture and a fundamental reorganisation of the way an enterprise operates. KM practitioners readily accept the centrality of cultural issues in KM, yet few businesses set out to actively manage the change that effective knowledge management entails. As so many firms have learnt to their cost, 'plug and play' is not a concept that applies here. Rather, the success of a KM programme depends on the adoption among knowledge workers of new behaviours and attitudes, a process that requires commitment, dedication and, above all, patience on behalf of those charged with leading the project. Companies that fail to address the change management issues that underpin KM will find knowledge management rapidly becomes an added, unwanted responsibility that employees are more than happy to ignore. Though few KM practitioners would dispute this, change management remains an aspect of their art that warrants far more attention than it currently receives. Admittedly, there are no set methods for dealing with change, but firms have a number of options open to them in how they tackle the human dynamics that affect KM implementation. The basic tenet of knowledge management is that knowledge is valuable and is required to do work and that the way knowledge workers acquire and share knowledge can be improved. Improvement in turn implies change; as such, managing change should enable improvement. Change management focuses on the human dynamics that accompany organisational change. Its objective is to drive a sustainable transformation in the way people work and interact. 'Ultimately, the success of a KM programme is linked to the level of adoption of new behaviours and attitudes among knowledge creators and users', he continues. Therefore, a well-thought-out change management strategy is instrumental to the success of any KM initiative. Without addressing the human dynamics, a KM programme will fail to deliver its projected results.

5.27   KM Measurement

KM measures should act as a dashboard to help the organisation understand where to make changes in KM initiative. KM measurement techniques include benchmarking, the balanced scorecard and the house of quality matrix. However, in the assessment framework of knowledge management, the following are vital:

5.27.1   *KM Metrics*

Metrics, also known as 'measures' or 'key performance indicators' (KPIs), are simply a tool for assessing the impact of a particular project or activity. While these are often numeric in nature ('improve sales by 20%'), they can also be qualitative ('improve staff satisfaction levels'). In either case, metrics provide clear and tangible goals for a project and criteria for project success. With the far-reaching impact of both knowledge management (KM) and content management system (CMS) projects, the use of metrics is even more important. Metrics are a concrete way of defining what a knowledge management or content management project will achieve and whether it met those goals.

Metrics cannot exist in isolation of business objectives. The first step is therefore to determine the goals for the project. Then key business metrics that are directly related to these goals need to be identified. The most powerful business metrics are those that directly measure the desired business outcomes. The implementation metrics that relate to measuring the success of implementing the content management or knowledge management system also need to be established. This comprises of system usage, number of users, information quality, user feedback, maintenance cost, staff efficiency and transaction costs. While many content management and knowledge management systems are targeted at internal staff, they can nonetheless have a considerable impact on the organisation's customers. Customer service metrics range from product sales, lead conversion, customer satisfaction, call handling time, transactions processed, etc. Knowledge management projects may endeavour to change the broader culture within an organisation, to match strategic needs or external factors. There are a number of metrics that can be used to measure the impact on staff culture like success stories, staff morale, anecdotes, satisfaction and staff learning.

Measures must be taken before a project is initiated, if metrics are to be fully effective. Only then can a clear 'before and after' comparison be made. By making these pre-project measures, however, it becomes possible to quantify the success of the project the moment the first metrics are assessed. As this is often immediately after implementation, this makes for rapid and tangible feedback, while the issues are still relevant. The chosen metrics should be reassessed every six to twelve months, to determine whether they are still effective and appropriate. It will often be necessary to drop some metrics or establish some new measures. This is not a bad thing and is often just a reflection of the changing business environment.

5.27.2   Return on Investment

It is important to note that the KM is one of the several initiatives that an organisation may take in this fast-changing world like TQM, new product investments and new market channels. Managing knowledge must prove to be beneficial to an organisation and thus stand on its own merit so that it is worth to be continued. In an organisation, there are competing uses of the limited resources by other initiatives. So there must be a clear analysis of the return on the investments being made and whether sufficient resources to aid those investment exist or not.

The KM initiative and its architecture shall be such that the efforts can be integrated into other programmes and initiatives of the organisation which will reduce the incremental costs directly attributed to the KM initiative while keeping the benefits intact. Also the benefits of KM will get a boost by the contributions being made by the other initiatives. ROI takes time to gather due to the complexity of understanding the ways in which people, process, content and technology affect knowledge sharing and, subsequently, the business. Many organisations embark on the KM journey since understanding, sharing and reusing knowledge seems a good business proposition. However, it requires great effort and investment to align people with tools, content and processes that facilitate knowledge flow, so while taking such an initiative, its feasibility must be considered as the resources for any organisations are limited and need to be effectively utilised. Management needs to assess the value of each investment and compare it against the potential gains from other initiatives or change efforts. Analysing ROI for KM is important to realise its contribution towards the productivity and performance of the organisation. Storytelling and anecdotal success stories are some of the ways to show the value of the investments made in KM.

In the following bell curve diagram (by the American Productivity and Quality Center) shows the importance of explicit KM activity measures over time. This has helped to identify KM initiative measurement approaches according to the relevance of activities in different phases of the implementation (Fig. 5.9).

5.28   KM Tools

As far as KM is concerned, tools and technology are inseparable as both of them together add speed and efficiency in driving KMS operations. The major KM tools are as follows:

1. *ICT*: It is defined as a set of technologies and tools to communicate, create, disseminate, store and manage information and knowledge. These technologies include computers, networks, the Internet, mobile devices, etc. They also

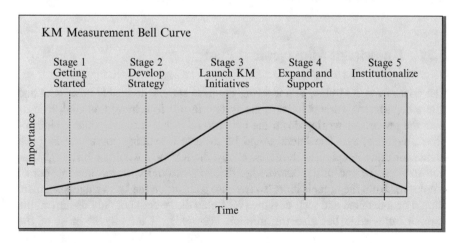

Fig. 5.9 KM measurement bell curve

include Web-based conferencing, audio and teleconferencing, etc. The potential capabilities of ICT are easy delivery access, presentation in any form, experimentation to test hypothesis, seamless communication of knowledge in any form, etc.

2. *E-learning, blended learning and distance learning*: e-learning is very popular in higher education and corporate learning. It extensively uses the Internet, intranet and extranet on LAN and WAN. It uses the Internet platform for knowledge delivery, interaction, communication evaluation and feedback. Blended learning assumes that not all learning is achieved through e-learning. It is necessary to have expert or mentor interaction and intervention and is largely decided by the characteristics of the learner and the subject itself. This method combines classroom interaction with e-learning and uses print material and online material for all steps of learning. Distance learning is chosen where reach and access is a problem, further compounded by cost of delivery.

3. *Unified communication technology*: UCT improves efficiency and effectiveness of supply chains by enabling stakeholders to collaborate even when located anywhere in different time zones. Different functional teams and experts, along with external supply chain and logistics partners, can share data and information to act fast. With integration of UC capabilities into business processes, employees can easily and quickly find the right person and communicate through the software applications and business processes currently in use.

4. *Communication-enabled business processes*: CEBP is about improving the processes and systems that drive business. A CEBP solution would be implemented to automate all communications in claims process with embedded UC services, like reminders, alerts, notifications and escalations. Quantifiable business results of such automation are decreased cycle times, greater HR productivity, enabling them to spend more time adjusting claims rather than fielding calls on the claims, increased revenue and customer satisfaction and retention.

There are other tools as well which can be effectively used for KM like WiMAX technology, data warehouse and data mining, search engines, knowledge portals, etc.

5.29   Knowledge Management Plan

The goal of any KM plan, *in a working progress state*, is to enable its present and future department workers to learn the system faster. To achieve this goal, it starts with the process of writing down the policies, mission statement, values, philosophies, methods, etc. The intent would be to have a learning curve that is much higher for future employees. Needless to say, the KM plan would help an organisation to preserve and utilise knowledge that may go out with the moving out of employees from the organisation. To exercise a check on the knowledge loss, there should be a detailed and well-designed KM plan that would not only document the explicit knowledge but also the implicit knowledge. Currently in most of the organisations, knowledge is passed down in a hierarchical fashion. However, much information and knowledge is lost in this manner. In most of the organisations, new employees are expected to perform without any formal introduction about the company's vision, mission, goal, objectives, history, culture, etc. This lack of knowledge about the organisation does not help the employees in the long run as the knowledge about the organisation would not only give them an idea about the needs of the organisation but also facilitate their work and thus help in achieving the organisation's targets. An example of an organisation like NASA would tell us about the utility and usefulness of having a good KM plan. NASA's knowledge, its intellectual capital, is the agency's primary, sustainable source of competitive advantage. This knowledge is a fluid mix of experience and know-how that allows NASA employees to strive for and achieve the improbable day after day. Those companies whose cultures promote sharing of knowledge and individual learning have high employee retention, attract high-quality employees and have a workforce that focuses on fixing the problem rather than fixing the blame. They have many of the key ingredients to making knowledge management succeed—a highly intelligent workforce, a need to learn in order to succeed and some solid, technical infrastructure.

NASA adopted a number of methods for different departments. For example, for the projects working under the Provide Aerospace Products and Capabilities (PAPAC), the process will clearly benefit from a set of collaborative tools for virtual teams and communities to share information. Such a capability would need to incorporate into the tools the rules related to working with external partners.

5.30 KM Risk Plan (Challenges and Issues)

KM has its own share of challenges and issues. As opposed to the popular belief that KM is a technology-only-based initiative, it also involves a lot of social and cultural issues. Dealing with these cultural and social issues while devising a good KM strategy is one of the serious challenges that an organisation's KM team has to deal with. The major problems that occur in KM usually result because companies ignore the people and cultural issues. In an environment where individual knowledge is considered to be very important and useful, it is critical to have an established culture that recognises tacit knowledge and encourages employees to share it among each other.

Challenges and issues with KM are as follows:

1. Unlike the popular belief, KM is not only a technology-based initiative. It should focus on people and cultural issues which in itself is a very tough task. KM decisions should primarily be based on people, business, existing knowledge, etc. and to assimilate all these components is not a simple job.
2. A KM plan should have a perfect alignment with the business goal and objectives, as a KM plan which is not in sync with the wider business goals cannot persist for a longer duration.

3. The most important issue is that knowledge is not static and can get stale fast. Hence, KM programme should be constantly updated, amended and deleted, if needed.
4. Further, the relevance of knowledge required also changes with the change in the prevailing business environment, existing employees, current organisational situation, etc. This means that KM has no ending point and it is a constantly evolving practice.
5. Companies need to be vigilant for information overload. Quantity rarely equals quality, and KM is no exception. Indeed, the point of a KM programme is to identify and disseminate knowledge gems from a sea of information.
6. The type of organisational culture often proves to be a KM barrier. This profile needs to be assessed and characterised in order to allow proactive steps to be taken.
7. The paradox of the value of an intellectual or knowledge asset is one of the major issues facing KM today. Human structural and customer capital would need to be codified to some extent, and their sharing promoted actively throughout the organisation.

5.31   Knowledge Management Audit

Almost all the organisations and entities are well acquainted with the management of different sectors like finance, marketing, operations, sales, supply chain, etc. However, this does not ensure that it would be easier for them to win in the ever-changing, dynamic and highly competitive markets that have emerged these days. Thus, the mantra to success today undoubtedly is innovation and invention. Thinking out of the box and something different from traditional ways is the need of the hour. The knowledge existing in the organisation, along with the knowledge that had existed, is very important to enable such fresh and innovative initiatives to flourish, further leading to the creation of new knowledge. This has become a very critical issue for the organisations today as far as KM strategy is concerned. However, in recent years, it has been found out that the key to success would be the efficient management of knowledge, external and internal, tacit and explicit, so that it can be ploughed back in the organisation to devise a new set of knowledge, leading to newer innovations. It thus helps them to create values to the organisations.

A knowledge audit methodology with emphasis on organisational core processes applied in a cyclical manner would be an efficient strategy to audit the key knowledge within an organisation.

By definition, a knowledge audit is an assessment of the way knowledge processes meet an organisation's knowledge goals. It is a means to understand the processes that constitute the activities of a knowledge worker and see how well they address the 'knowledge goals' of the organisation. Liebowitz defines a knowledge audit as a tool that assets potential stores of knowledge. It is the first part of any KM strategy.

The knowledge audit is the first step of a KM initiative. It is used to provide a sound insight into the ways an organisation is functioning. The knowledge audit examines knowledge sources and uses how and why knowledge is acquired, accessed, disseminated, shared and used. It gives an insight regarding the organisation's readiness to become knowledge centred. It also intends to assess the levels of knowledge that is being used in the organisation, knowledge interchange and dissemination, knowledge management propensity within the enterprise, identification and analysis of knowledge management opportunities, isolation of potential problem areas and an evaluation of the perceived value in knowledge within the enterprise.

5.32 Knowledge Audit Methodology

As already stated, knowledge audit needs to be done in the initial stages of the KM programme. It should broadly identify the knowledge requirements of all the methods and processes in the organisation that are largely dependent on the intellectual repository and that target the underlying business goals. The main function of KM audit is to identify and locate the knowledge sources that are capable of fulfilling these knowledge requirements and the high-level business process steps where that knowledge can be applied. It is generally advisable to start the audit with a small group or team or a business unit. It can be done by dividing it in two tasks, which are not necessarily dependent upon each other. The first one, popularly known as knowledge mapping, intends to locate the repositories of knowledge throughout the organisation. This effort is primarily technical and helps the organisation in the preparation of a knowledge database. The knowledge mapping is a very simple process that takes the inventory of what people have written or entered in the information systems, besides identifying the sources that employees use for information from outside. However, finding and collating all such data is a time-consuming task, but is conceptually easy comparatively. The second category of audit task helps in capturing the patterns of flow of knowledge in the organisation. While there seem to be several ways of conducting a knowledge audit, in general, knowledge audits consists of the identification of knowledge needs through questionnaires, interviews and focus group discussions; the development of a knowledge inventory is mainly concerned with the available knowledge types; maintenance and storage of knowledge; its location, mode and purpose of the usage of knowledge; its relevance; knowledge map creation; knowledge flow analysis; and finally a detailed audit report. However, auditing is a very subjective task as it can be done in different ways by different organisations. Of late, there has been a demand of a basic rough outline to facilitate the conduct of auditing process smoothly so as to realise the potential benefits of the organisation.

5.33   KM: The Way Ahead

KM, as it used to be initially, has been only the means to discover the ultimate truths and organise the world according to rational people using the clever code. However, in this modern era of knowledge economy, KM has many diverse and varied roles to play. The knowledge management in the postmodern era should not be restricted to management only, as it is more of a top-down approach and implies external control of some definable source. It should aim at leveraging people and knowledge, rather than just managing them. Therefore, modern KM should aim at fostering human intelligence and interaction rather than trying to replace them. Further, the transformation of R&D from its 'product-centric' approach to a 'knowledge-centric future' also paves a new way for the management of knowledge in organisations. In addition, their focus on 'discontinuous' and 'fusion' innovation promises to lead the way for the industry. All these upcoming trends signal the tearing down of R&D's overly centralised and compartmentalised profile in most of the firms, offering strong support for the view that innovation should be structured as a distributed, whole-firm social process, rather than an administrative one.

5.34   Summary

The chapter gives us a deep insight about the importance of KM to an organisation. Apparently, it seems to be the only exploitation of knowledge and information within an organisation. However, in its deeper sense, it is a way of building the foundation for improved business advantages and strengthens the capabilities of a sustained future. The more the KM in an organisation is embedded in its culture, practices and processes, the more successful it ought to be. However, in view of developing a good KM system, it should not be forgotten that the real assets of an organisation are its employees and KM is not only about managing knowledge and people; it is more about leveraging people and knowledge and fostering culture, interactions and sharing of knowledge and ideas. KM is not a way out to develop knowledge in an organisation; it is rather a way to get into an organisation and create a source of knowledge from the existing sea of unorganised knowledge and information. Thus, KM is a catalytic process that performs a plethora of functions— ranging right from impacting individual behaviours, collaborative knowledge sharing within groups and organisational dimensions such as culture. The true value of KM is achieved when it moves from the status quo to beyond rhetoric in the realm of the organisation's business needs to meet its long-term goals and to thrive in an economy which is knowledge driven.

Chapter 6
KM Metrics and KM Audit

6.1 KM Metrics

Knowledge management (KM) would be aptly defined as a logical means of maintaining and keeping the past wisdom recorded and available for use in the future or for setting a standard which can either be modified or improvised for application in all fields of study so that one does not have to begin from scratch every time. Thus it can be seen as a fast track way of learning the basics for a new entrant in any field that requires a sequence set of instructions before one starts experimenting with it. This could be in two forms:

- As explicit or something which is known and present in records.
- As tacit or the accumulated wisdom that one gathers by experience and expertise by working on it for a considerable time. These may be just ideas as well which can later be materialised by another person.

Knowledge management which is a relatively new field that tries to capture and capitalise on this tacit knowledge and convert it to application so that it can finally be standardised and reused with modifications and improvisation includes numerous subverticals in it. These can be depicted as (Fig. 6.1):

According to the American Productivity and Quality Centre, knowledge management can be put into words as:

A systematic process of connecting people to people and people to knowledge and information they need to effectively act and create new knowledge.

© Springer International Publishing Switzerland 2016
S. Mohapatra et al., *Designing Knowledge Management-Enabled Business Strategies*, Management for Professionals, DOI 10.1007/978-3-319-33894-1_6

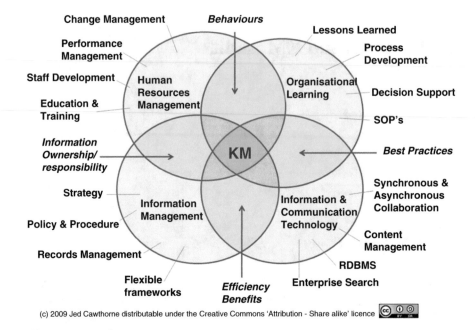

Fig. 6.1 KM metrics

The two types of knowledge that is supposed to be residing within an organisation are:

- Explicit which includes a standard and formal set of instruction or codes that one needs to follow while working on a specific task. These could be further extended for reports, copyrights and so on.
- Tacit includes the knowledge that is yet to be documented since it is in a very informal manner with people working in the organisation and those who have spent some time working in any specific field. This can be equated with colloquial wisdom of the domain, something which is very useful but gets lost generation by generation in the absence of proper documentation.

Today an organisation is known more by the intellectual capital that it possesses in comparison to its physical assets. The intellectual assets include patents, ideas, innovations in technologies and designs that have made firms like Bose, 3M, Boeing, etc. far ahead of their competitors. The challenge for these organisations however remains in documenting these developments in a manner that it reduces the lead time for any new entrant by standardisation and also gives it a clear guideline or direction for the scope of improvement so that the efforts are non-repetitive and time saving.

Hence the knowledge captured and which is in a continuous mode of being improved further needs to be categorised at every step. This can be effectively done on four fronts.

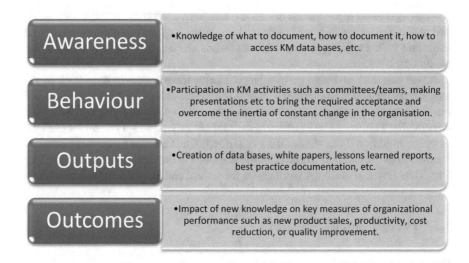

For example, Ford as a company laid stress on the end results, this could be either the output or the outcome. For ford it has never been the number of documentation, but it was always the number of modifications made and implemented in the production design. Thus we could say that Ford Motor Company always encouraged innovation, but for them the measure was not in reports but the bottom line outcomes.

Ford being an old corporate giant and having incorporated the knowledge management long back in their regular processes has already incorporated the factors of awareness and behaviour in their organisation culture itself. Hence even when a new employee joins the firm, he is given an induction for the same so that the lead time for output generation is reduced considerably.

6.2 Measures for Outcomes

	Methods for measuring	Metrics
Learning and innovation	– Improvement programmes	1. No. of improvement/innovation ideas
	– Meetings, workshop	2. No. of new services/products ideas
		3. No. of internal meetings (for sharing knowledge and experiences)
Capability	– Questionnaire	1. No. of trainings per year/per employee
	– Individual assessment	2. The analysis results of employee satisfaction surveys
		3. KPIs for individuals
Productivity, profitability, growth	– Business analysis	1. ROI
		2. Image/reputation
	– Market research	3. Market share/index
		4. Revenue for new products

Source: http://thinkingshift.wordpress.com/category/knowledge-management/

Thus ford as an organisation has a very integrative structure which is decentralised at the departmental level but centralised at the organisational level so that no effort that yields result goes wasted and reaches all the units for application at the production level.

It was essential to design metrics since it helps the mangers to keep track of the progress and also to measure it in a quantifiable term. This could further be used as a tool for appraisal system if maintained for the entire project team basis and even individually. This would be an added benefit for identifying an apt person for each job on the organisation if one studies the trend of individual metrics for the work done by a person in case of an initial time if he/she is rotated in the various departments of the firm. Thus it would make the work easier for reward and incentive system as well.

6.3   Performance Indicators

The required qualities and characteristics on performance indicators are:

- *Countenance* that would include the usage and sharing of rights of a person's work and modification of a paper that could be shared or made public only with his prior consent.
- *Operational nature*, which is the ability to assign correct value to the trait which is being measured. This might be a little easier for the quantitative data as financial figures but a little difficult for abstract qualitative aspects.
- *Significance*, which indicates the place that the information attains in the organisation that is how far it has been used and reused over a particular interval of time (Fig. 6.2).

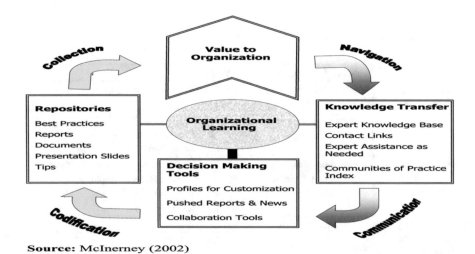

Source: McInerney (2002)

Fig. 6.2 Organisational learning chart

The KM indicators can be further classified as:

- Process indicator also known as lead or effort measures which signify the new initiatives of the organisation.
- Result indicator also known as lag measures. These are concerned with the strategic and operational achievements that have been documented in the metrics and are then used to decide the future course of action through either a gap analysis or trend analysis and so on.

These indicators are used in the resolution and prevision of problems. The performance output indicators reflect the result of a process.

Price Waterhouse lays stress on building a collection of well-balanced indicators that incorporates and complements all the processes of knowledge management in the work. This should have not only the numerical or quantitative data but also the qualitative aspect that should be equilibrated along with financial and nonfinancial ones. They recommend that all these need to be seen as one integrated entity and their relationship be established in a manner so that we get a crystal clear idea of what is affecting where, why and by how much (Figs. 6.3 and 6.4).

However there are some doubts about the KM programmes being a failure as well. This is mainly due to the inefficiency while developing the critical success factors in the KM metrics:

- Knowledge management usually handles assets that are intangible and include structural capital, human resource, etc. for the organisation.
- These are very complex and are hence referred to as the 'hard' metrics. Many people thus prefer the 'soft' measures which go by the norms of using anecdotes or organisational narratives, to explain the usefulness and utility of their organisation's KM system.

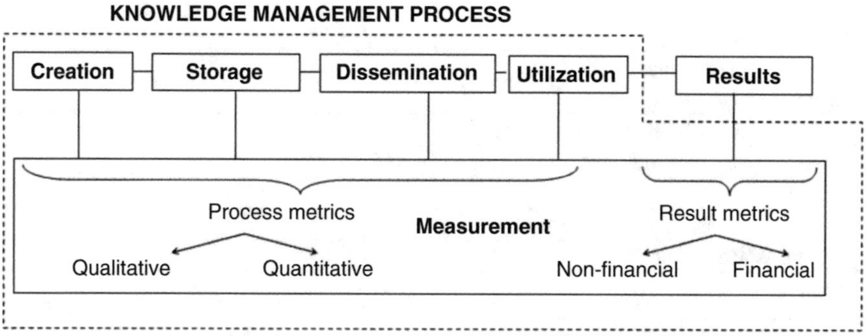

Fig. 6.3 Knowledge management process

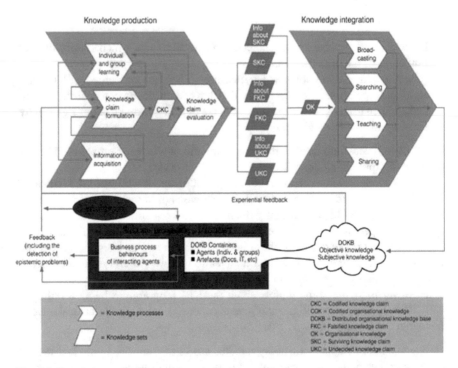

Fig. 6.4 Knowledge production and knowledge integration. *Source*: http://www.ikmagazine.com/xq/asp/txtSearch.CoP/exactphrase.1/sid.0/articleid.CFF1EF12-3B9E-4A62-9091-010C44228A81/qx/display.htm

6.4   Summary of KM Performance Measures: Personnel and Training

Business objectives
Alluring and retaining the talent pool
Increase chances of learning
Special attention on enhancing the quality
Necessary steps: measures essentially required for KM initiatives
Outcome
Reusability analysis through survey
As stories and anecdotes
Structures
Response time or lag period
Number of people who hit the site and finally download
Time spent on any page or section of the material available
Rate or use frequency
Navigation path for the search being followed

Calls mane to the helpline			
Percentage of employees using or availing the facility			
KM steps	*Structure*	*Outcome*	*Result*
Portal	Accuracy and recall value	Reduce cost of printed newsletters	Reduction in induction period for a new employee
For HR	Personalising the trait	More time for information collection	– Standardisation of instruction manual
Functions	Rate of search		
	Frequency of it being the home page		
Communities of practice	How frequently contributions have been made and they have been upgraded	Number of complaints registered	– Reduced turnover
–Common interest group that meets informally or formally for exchange of views and ideas	How many users have become the contributing members	Attrition rate for the organisation or the project	Increase in acceptance of employment offers
		Number of employees who have joined other firms	Higher work efficiency and satisfaction level of workforce

For professionals, metrics are indicators of what works and what does not. In fact, going by an example measuring business performance is of utmost importance when it comes to managing the accounts of a firm. Similarly in KM, obtaining funds for executing the KM initiative to measure performance in each vertical for providing the periodic targets and finally gaining unbiased feedback for deciding the future course of action and the changes to be made is very crucial. These help not only in evaluation of the initial investment decision but very importantly in developing the benchmarks for the future comparisons. One should be careful with the fact that measurement is usually laden with mistakes that could either be subjective in terms of how and who evaluates them or objective in case the metrics designed does not cover all the critical success factors of the project or assignment. Importance and ease of measure may not always go hand in hand for KM strategy implementation.

The following structure has been proposed while trying to formulate the metrics for implementing the KM cycle in any organisation.

6.5 Structure

1. Retrieve and display strategic goals and objectives, tactical goals and objectives and plans for knowledge discovery from outputs.

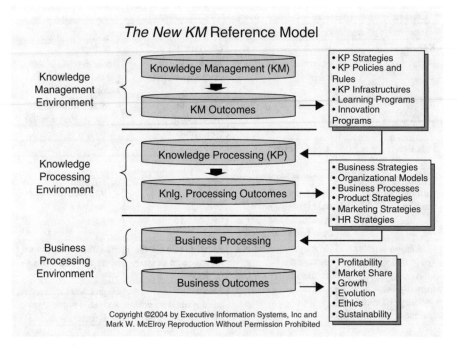

Fig. 6.5 New KM reference model. *Source*: http://kmci.org/alllifeisproblemsolving/archives/
why-dont-we-write-much-about-km-metrics/

2. Select entity objects representing business domains to be mined for new
   knowledge.
3. Sample data.
4. Explore data and clean for modelling and recode to transform data.
5. Reduce data.
6. Select variables for modelling.
7. Transform variables.
8. Perform measurement modelling.
9. Estimate models.
10. Validate models (Fig. 6.5).

However this is not the end. These further need to be done minutely and the various subtasks of the validate models task that need to be taken into consideration.
These are:

1. Evaluate hierarchy of the factors that need to be considered in any specific
   process.

2. Arrange as per highest level of validity vis-a-vis their relativity to one another.
3. Note the highest number of attribute for each factor.
4. Arrange attributes of the highest relativity as in (3) above, as related to each other and the factor under which they have been mentioned.
5. These factors and attributes should be in compliance with the business objectives and the goals mentioned in (3).
6. In case it is possible and deemed necessary, the attributes should further be subclassified.
7. Repeat (5) and (6) till one reaches the lowest level of classification for goals to be fulfilled and undertaken by the lowest level of staff involved in contributing to the assigned project or assignment.
8. This arrangement of order should then be recorded.
9. This validity-related attributes and then their order must be evaluated once again on an organisational level so that final changes could be made before making it uniform for the execution across the organisation.
10. Insert these attributes in software.
11. Compare results obtained by analysis of factors as clustering and so on to reach the goals specified in the beginning in the given timeframe.

6.6 Knowledge Audit and KM Audit

Any audit aims to check the compliance of the stipulated processes. KM audit is no different; it checks whether the proposed KM activities in the KM plan are carried out. Knowledge audit as per Lauer and Tanniru is the understanding of the processes that constitute the activities of a worker in achieving his knowledge goals for the organisation. The most important starting point of a KM initiative is the knowledge audit. This can be seen as the KM audit before the KM is implemented. A knowledge audit helps in assessing the knowledge health of the organisation. Through this audit, one can baseline the knowledge levels of the organisation. The knowledge audit is a discovery, verification, and validation tool, providing fact-finding, analysis, interpretation and reports.

Outputs and outcomes that are expected out of the KM process are:

• Current rate at which knowledge is being used in the organisation.
• How is knowledge accumulated/trickled within the organisation.
• Gaps and opportunities in knowledge management of an organisation or if at all need to be started or completely restructured.
• Identifying the dark spots or problematic areas of concern.
• Assessment of tacit and perceived knowledge of the firm.

6.7   KM Audit Process

The main aim of KM audit process is to identify all the processes that use high intellectual knowledge and are critical to business objectives. Once these processes are identified, specific action items can be incorporated in KM activities to reduce the dependence on the tacit knowledge. Following is the KM audit process that can be applied to various kinds of organisations (Fig. 6.6).

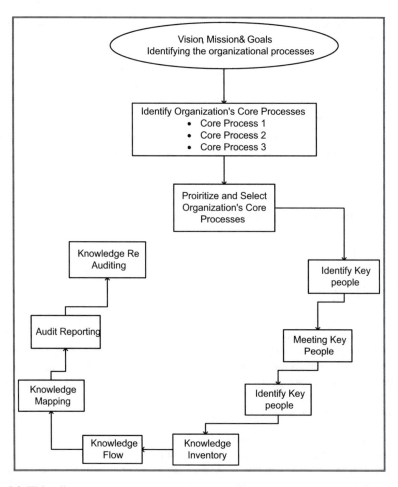

Fig. 6.6 KM audit process

6.8   Steps of KM Audit Process

Steps of KM Audit Process

Stage 1
- Acquire organizational strategic information and identify organizational processes.
- Identify mission, vision and organizational objectives considering the environment, culture and traditions.

Stage 2
- Identify organization's core processes and establish measurement criteria
- Identify organization's core processes that contain useful knowledge to be managed; measure the performance of the knowledge processes within core processes.

Stage 3
- Prioritize and select organization's core processes
- Prioritize and select organization's core processes according to the criteria defined at second stage.

Stage 4
- Identify key people
- Identify the key people who participate in the selected core processes.

Stage 5
- Meeting with key people.
- Give information to key people about knowledge audit and knowledge management processes.

Stage 6
- Obtaining knowledge inventory
- Locate and obtain existing knowledge assets within the organization.

Stage 7
- Analyzing knowledge flow
- Analyze how knowledge within the organization flows.

Stage 8
- Knowledge mapping
- Visually represent organizational knowledge. This map includes knowledge inventory and knowledge flow within the organization.

Stage 9
- Knowledge Audit Reporting
- Give the organizational managers' outcome of knowledge audit. The results form valuable information for strategic planning. This report gives the final validation and justification for the short, medium and long term KM strategy and investment.

Stage 10
- Continuous Knowledge Re-auditing
- Analyze and select the rest of the core processes to complete the knowledge audit
- Update any changes of the knowledge inventory, knowledge flow, knowledge map, and the knowledge processes.

6.9   Conclusion

Any KM implementation will not be successful as long as it has business-driven KM metrics, critical success factors built into it. KM audit identifies the gap and helps in achieving the business goals faster and in an efficient process. KM metrics are crucial as they are the parameters on which the entire knowledge management decisions are taken. Hence choosing the right parameters is of prime importance. The tracking of the metrics are done through the audit process. The audit process helps in assessing how much we are aligned to the business goals and where to improve the process. Hence both metrics and audit are very critical to the success of KM implementation.

6.10   Summary

Metrics is a word that is derived from metre—a Latin word which means to measure. If we want to improve, then we should be able to measure the parameter. In organisations, it is important to measure effectiveness of knowledge management strategy. This will help to align business objectives and outcome of KM strategy. The alignment will help to implement and reap business benefits. KM audit is used to find compliance to KM strategy and provide feedback periodically so that not only compliance to KM strategy improves, but also strategy is refined for better alignment.

Chapter 7
Implementing KM Strategy

7.1 Introduction

An African proverb states that 'Knowledge is the only treasure you can give entirely without running short of it'.

The way in which to preserve this knowledge and pass it on to others is through knowledge management (KM). KM is about managing knowledge—capturing, creating, reviewing, coding and sharing. It is about communities that have an interest in a particular subject, and they keep the knowledge alive, build on it, use it for their own purposes and later share it, thus creating best practices and processes.

The following figure shows the KM cycle (Fig. 7.1).

The above cycle gives a brief pictorial representation of the KM cycle. We have included the implementation phases in this cycle since KM implementation would be the topic of discussion in this chapter.

The above diagram indicates that tacit knowledge is present within an individual or a group. This tacit knowledge, unless captured, would walk out of the organisation in case the individual or the group decides to leave the organisation. This knowledge is then captured by say a community of practice (COP) and is then assessed (peer review). Knowledge creation refers to the addition of new knowledge within an organisation which was not present earlier. The assessment/review is also done by a subject matter expert (SME). The knowledge is then codified based on a certain taxonomy or nomenclature, and it identifies the key attributes of the document. Thus, half cycle completes in the conversion of the initial tacit knowledge to explicit knowledge.

The explicit knowledge is then shared. This requires systems for the creation and maintenance of knowledge repositories and to cultivate and facilitate the sharing of knowledge and organisational learning. Access to the knowledge implies that there are certain rights and privileges to certain groups of users over certain documents. Not everyone will be able to access the entire knowledge repository. It would depend on the persons' roles and responsibilities in the organisation.

© Springer International Publishing Switzerland 2016
S. Mohapatra et al., *Designing Knowledge Management-Enabled Business Strategies*, Management for Professionals, DOI 10.1007/978-3-319-33894-1_7

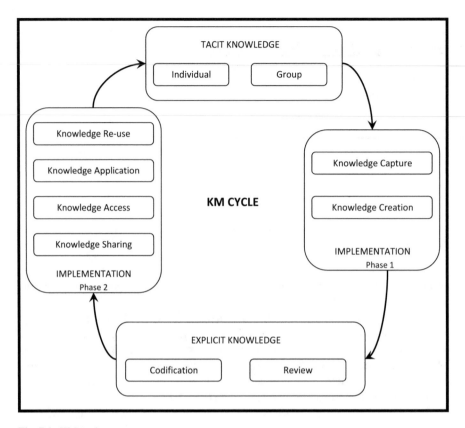

Fig. 7.1 KM cycle

The people of the organisation would then apply this knowledge in their respective areas of work, thus saving time and improving performance levels. This knowledge would be used and reused and changed as per the specific requirements. Thus, once again explicit knowledge would lead to more tacit knowledge based on the implementer's interpretation and use of the explicit knowledge and the cycle would repeat.

In the following sections, we would look at KM implementation in details with specific focus on the various frameworks of implementation and steps and gaps in implementation identified in literature so far.

7.2   KM Implementation Frameworks

In this section, we have tried to cover some basic KM implementation frameworks. We have tried to show frameworks for corporate as well as development organisations.

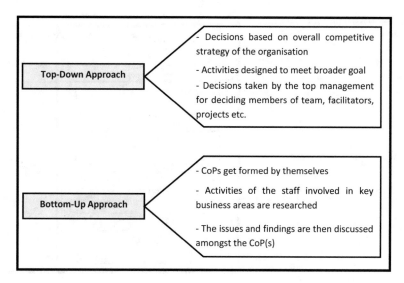

Fig. 7.2   Top-down and bottom-up approaches

Typically in an organisation, KM implementation can be either top-down or bottom-up approach. The following figure gives the information about these two approaches (Fig. 7.2).

Depending on the competitive strategy, type of business operations, business goals, scale of operations, attrition rates, functional goals, etc., an organisation can decide to go for KM. The KM strategy has to be first defined based on any of the above two approaches and then needs to be implemented. The KM strategy also needs to define the outcome that needs to be achieved with KM implementation. We have tried to cover implementation frameworks for a corporate and an NGO.

7.3   MAKE Plan for KM

MAKE stands for most admired knowledge enterprises. MAKE awards are conferred upon organisations that fulfil the following criteria:

- Creating a corporate knowledge-driven culture
- Developing knowledge workers through senior management leadership
- Innovation
- Maximising enterprise intellectual capital
- Creating an environment for collaborative knowledge sharing
- Organisational learning
- Delivering value based on stakeholder knowledge
- Transforming enterprise knowledge into stakeholder value

Fig. 7.3 MAKE plan for
organisations

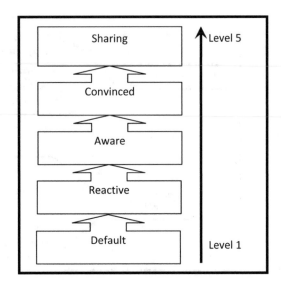

These awards are given to organisations in order to enable development of
knowledge and skills in the industry. It also includes inculcating the sense and need
for knowledge management in organisations. The following figure shows the
MAKE plan for KM (Fig. 7.3).

Sharing is the highest level that an organisation can achieve in the MAKE framework. At this level, ROI drives the decision-making, and there are set processes in
place to leverage new ideas.

7.4 KM Implementation Framework

The following figure shows the KM implementation framework that can be used in
corporate organisations (Fig. 7.4).

The figure above shows that the first activity in KM implementation would be to
form the KM teams based on the business goals. Networking, collaborating and
integrating are very essential in order to form an effective team and identifying the
areas for KM intervention.

A detailed KM implementation plan needs to be defined based on the KM strategies that get decided. In order to successfully implement the KM plan, proper infrastructure is needed. This infrastructure needs to be developed.

The KM implementation should first be done on a pilot basis for a particular
functional area or a business unit. The PDCA cycle can be used in this phase to
continuously improve the process. It will also help to plan and analyse carefully and
to identify gaps and issues within the process. The next step would be to implement
KM either on an organisational level or a business level as may be the case. This will
depend on the success and analysis of the pilot phase.

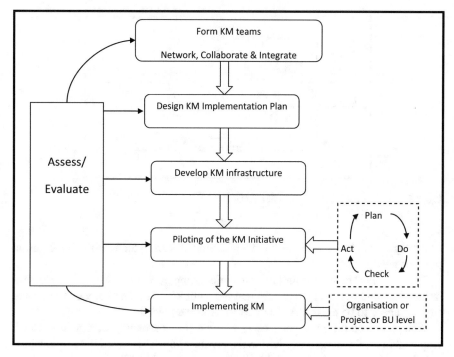

Fig. 7.4   KM implementation framework for corporates

At each level of the framework, continuous assessment and evaluation is very essential. This will help to identify gaps and issues in the implementation in a timely and effective manner thus enabling changes immediately leading to reduced costs and time.

7.5   KM Implementation Framework for a Non-profit Organisation

Knowledge management is as important in the non-profit sector as it is in the for-profit sector. One of the major reasons for the failures of projects and difficulties and lack of success in replicating successful models is the lack of KM initiatives. In this section, we will try to develop a framework for KM implementation in a non-profit organisation.

The following figure shows the Leavitt's model of organisational change (Fig. 7.5).

This model basically states that in a non-profit organisation, there must be balance and coordination between the four mentioned subsystems, namely, technology, actors, structures and tasks, for the success of the KM initiative.

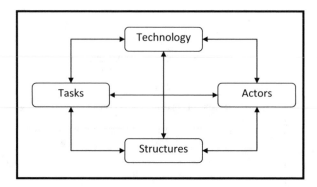

Fig. 7.5 Leavitt's model for organisational change

Though this is a generic model, we will look at it from the KM implementation point of view.

Technology refers to the tools and technical support required for KM implementation. An NPO needs to make sure that the right kind of technology and technical support needs to be present for the KM to be implemented.

Actors can be the managers, users or even the community of practice (COP) responsible for the initiative of KM in the organisation. Their roles and responsibilities should be well defined, and the KM practices should be linked to performance in order to motivate and sustain the KM initiative.

An NPO should be properly structured in order to enable KM. The structure of an organisation will help determine the area(s) of KM implementation and in steps such as piloting and analysing the impact of the KM initiative.

The tasks, which include the goals and deliverables, are essential to design and develop the KM implementation process. The KM strategy must be based on the working of the organisation and the areas of intervention.

All in all, the most important criterion is the coordination and the balance of all the different subsystems. In case the processes are in place and well structured, the framework stated in Fig. 7.3 above can also be used for NPO's though the scale of operations would be lower.

In the next section, we will look at detailed steps for KM implementation in an organisation.

7.6   Steps for KM Implementation

The following figure shows the ten golden steps which need to be followed for ensuring proper implementation of KM. The steps have been explained in detail below (Fig. 7.6).

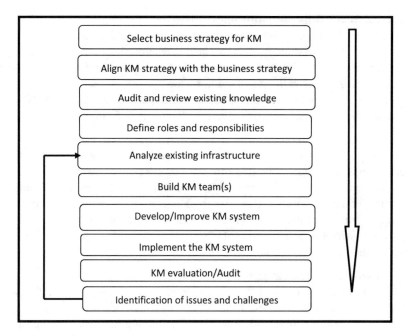

Fig. 7.6 Steps for KM implementation

Step 1: Select a Business Strategy for KM

This is the most elementary step to take for KM implementation. An organisation can have more than one strategy or business goal. Select the business goal for which KM needs to be implemented. It will give focus to the implementation and KM integration will be easier.

Step 2: Align KM Strategy with the Business Strategy

The KM strategy should be such that it should support the organisational goals and objectives. The organisation needs to analyse its internal environment. The internal environment has different processes and actors interrelated and interdependent with each other. The organisation needs to carefully examine the ongoing process and slowly integrate it with the organisational culture.

Step 3: Audit and Analyse Present Knowledge

Under this step, the present tacit knowledge available within the organisation needs to be checked and audited. This provides the information about the present knowledge flow of organisation and knowledge mapping can be done. This step gives opportunities for analysing the gaps, the opportunities and the problems within its business environment. This will give information about the current ongoing practices followed in the organisation.

Step 4: Analyse Existing Infrastructure

Analysing the existing infrastructure supports availability and feasibility for implementing the KM strategy. This will help in determining the requirements, gaps and potential available for the processing of KM in the organisation. Analysing the current infrastructure will give the information about the scale of investment required for fulfilling the gap and ensuring the proper functioning of the KM processes and in order to use present resources effectively and efficiently.

Step 5: Build KM Team(s)

Building KM teams is one of the important decisions that need to be taken for effective implementation of the KM strategy. The selection of members of KM team must be aligned with the business objectives and goal. If it is cross functional, proper mix of the different functional/business units must be maintained. Community of practice can also be the determinant of the members of the KM team. The number of the teams will depend on the size of the organisations. KM team consists of sponsors, facilitators and members.

Step 6: Define Roles and Responsibilities

Define roles and responsibilities clearly for each individual in a team for proper functioning of the KM activities. This will give each individual a clear set of activities. This will help in problem identification. By defining roles and responsibilities, conflicts can be minimised.

Step 7: Develop the KM System

Developing KM system is one of the important issues for proper implementation and functioning of its activities. The system should be such that the people should be able to integrate KM activities with their daily working activities, and it should support their working environment. The process of the KM system should be easy to use, and the decisions should be participatory in approach. The system should be built around the people and should help in their daily working. The matrices and different critical success factors (CSFs) should be defined, and this will entirely depend on the internal dynamics of the organisation. The matrices and CSFs are evolving processes and must be in line with business goals and objectives of the organisation.

Step 8: Implement the KM System

This is the ground part of the KM implementation process. KM can be implemented as part of an organisation change initiative. The main obstacle for implementing KM will be from the employees of the organisation. The working group will be resistant to change. Cautiously dealing with such obstacles is very critical to the KM system implementation. Employees need to be convinced that they will have the full support of the top management. Their performance indicators should also take KM work into consideration, i.e. the key performance indicators (KPIs) must be aligned with the KM strategy.

Step 9: KM Evaluation/Audit

KM is a continuous process and it requires continuous improvement and feedback. Under this step, organisation defines the period of evaluation. The success and the failure of the system will be checked on the basis of predefined matrices and CSFs level.

Who will perform evaluation/audit will mainly depend on the organisational culture and the internal dynamics of the organisation. Evaluation could be done by using questionnaires, interviews and focus group discussion.

Step 10: Identification of Issues and Challenges

Identifying the challenges and gaps is the most important function in KM strategy because KM is a continuously evolving process. It requires continuous reviewing. Here the main focus is not only identifying but also rectifying the problem faced due to the KM initiatives. It could be process improvement processes (PIPs) or it could be structure improvement processes. Decision will be taken by the group or the top management team and will depend upon the scale of the problem.

7.7   KM Implementation Roadmap

The following figure gives the KM implementation roadmap. The steps must be implemented and assessed as per the roadmap decided for implementation (Table 7.1).

The timeframe (weeks/months/years) will depend upon the organisational top management decisions. The timeframe will also depend upon the available infrastructures and resources. The area of work and expertise of the organisation will also be one of the determinants of the timeframe for KM implementation. The roadmap is usually shown for a 3–5-year strategy. The milestones to be achieved during each phase must be well defined and documented.

7.8   Evaluation of KM Implementation

KM is evaluated on the basis of matrices and predefined critical success points by the organisation. The data are captured on the basis of matrices defined in the template. This could be understood by the following example.

Table 7.1 KM implementation roadmap

Activity/phases	Phase 1	Phase 2	Phase 3	Phase 4	Phase 5
1. Select a business strategy for KM					
2. Align KM strategy with the business strategy					
3. Audit and analyse present knowledge					
4. Analyse existing infrastructure					
5. Build KM team(s)					
6. Define roles and responsibilities					
7. Develop the KM system					
8. Implement the KM system					
9. KM evaluation/audit					
10. Identification of issues and challenges					

For example, consider an organisation ABC. The goals of the organisation are given below and are aligned with the vision and mission of the organisation.

Goals of organisation ABC:

- Increase market share.
- Reduce cost.
- Increase efficiency of the resources.

We need to realise that not all goals can be achieved by KM implementation. We have considered the goal of reducing costs in the following section.

In order to reduce costs, there will many tactical, functional and operational level strategies. These strategies are decided by the top management. Management will also decide the percentage of increase in performance as a direct result of KM initiative. This will determine the critical success factor for KM evaluation. Current level of performance will be treated as base for comparing the change.

The following are the ways in which KM can reduce costs:

- Easy availability of information reducing the work load and thus saving employees' time.
- Information reuse.
- Easy availability of success/failure of certain initiatives previously undertaken in the project.
- KM will also enable discussions and provide information about subject matter experts (SMEs).

The following table gives an example of a template that can be used to evaluate the KM implementation process. In this case, we have considered a project level KM implementation. Based on each employee's usage of the KM structures and initiatives, we would be able to evaluate the project level improvement.

Sl. no.	Employee name	Work for which KM used	% of reuse (information)	Time reduced	Improvements made	Comments

On the basis of above template, the data will be captured. After capturing the data, the computation will be done and costs will be recognised. Then evaluation of the result will be done whether it successfully achieved its result.

7.9   KM Maturity Model

The figure below gives an example of a maturity model for an organisation undertaking a KM implementation process. This model is something we found on the Internet and would be useful in measuring the organisation's KM strategy based on certain set parameters (Fig. 7.7).

Questions	Weight	Measure (monthly)					
		0	1	2	3	4	5
1 Are we effectively **capturing** new people/experts, project work, new learnings, new ideas, and insights?	15	2	5	8	10	12	
2 Are we effectively **storing** new people/experts, project work, new learnings, new ideas and insights?	5	2	3	4	4	4	
3 Are we effectively **sharing/applying** new people/experts, project work, new learnings, new ideas and insights?	15	0	5	8	11	12	
4 Are we effectively **collaborating and discussing** problems, issues, new learnings, new ideas and insights through a shared knowledge space?	30	0	5	10	15	20	
5 Are we effectively **harvesting** new skills, competences, new learnings, new ideas and insights?	20	0	2	7	12	15	
6 Are we effectively **developing and applying** best knowledge/products/services and best practices/methods/designs?	30	0	2	10	18	22	
7 Are we effectively **developing** best experts and **capitalising** on Communities of Practice?	50	0	5	15	20	25	
8 Are we effectively **measuring** the above KM activities?	25	0	8	10	12	15	
9 Are we effectively **maintaining** the system?	10	0	0	5	5	7	
Total	200	4	35	77	107	132	

Fig. 7.7 KM maturity model

7.10   Gaps in KM Implementation

In this section, we will look at some of the gaps that can arise in the KM implementation process.

Gap 1: Alignment of the KM Goals with the Business Strategy

- For effective KM, the goals need to be in line with the business strategy.
- Not all the business goals can be achieved with KM implementation.
- For organisational level KM implementation and for scaling up of the KM from project or BU level to organisational level, it is very essential that the KM goals are aligned with the business strategy.

Gap 2: Improper/Ineffective Plan for KM Implementation

- If the planning and the implementation team are not properly briefed about the reason for KM initiative being undertaken, the entire plan may be ineffective or improper leading to failure of the implementation.
- This also will depend on the first gap mentioned above.
- The KM plan should be such that it must take the long-term plans into perspective and must be able to be implemented on an organisational level.
- It must also take into account the available resources and infrastructure with the organisation.

Gap 3: Miscommunication About the KM Initiatives Between the Top Management and the Employees

- The employees must be made aware of the KM initiatives being undertaken by the organisation.
- Strategic management must be implemented in order to handle resistance towards change which might be displayed by the employees.
- Also, the top management must involve all the employees' right from the initial stages of implementation and must communicate openly and transparently with them in order to make the entire process a participatory one.
- Even in case of top-down approach, such communication is a must.

Gap 4: No Rewards and Recognition Schemes for KM Initiatives and No Alignment of KPIs with KM Strategy

- For KM implementation to be taken seriously, the KPIs of the employees must be aligned with the KM strategy.
- Employees must be encouraged to submit white papers and for constantly managing and updating the KM repositories.
- Proper R & R schemes must be put in place to motivate and encourage the employees to participate in the process.

Gap 5: Mismatch Between Set Goals and Outcomes Achieved

- In order to avoid the mismatch between the set goals and the final outcomes achieved, the KM implementation process must be audited and evaluated at each stage.
- Gaps must be identified and rectified immediately to save on costs and time.
- Metrics and CSFs must be defined and constantly monitored and recorded in order to identify discrepancies immediately.

While implementing KM, the above gaps need to be paid attention to. It is very essential that all these gaps are avoided for a smooth and effective KM initiative. As mentioned, it is very essential in the piloting phase so that it becomes easy to scale up to the organisational level.

7.11 Conclusion

KM implementation should be holistic in nature, and it should take each and every stakeholder into its consideration. It requires a lot of background work. The employees will be resistant to change in their current working procedures, and this resistance has to be strategically managed. KM implementation should bring change slowly and patiently, and if it is done too fast, the objective of implementing will not be achieved.

KM processes should be in line with the current business processes/operations, and KPIs must be aligned with KM strategy. KM work should be carefully integrated. People should feel their job is getting easier with the implementation and it should not be seen as a burden.

KM has supported many small and big organisations to achieve its organisational targets and goals. KM implementation process plays a very important role in the successful operations of KM functions and KM strategies in an organisation, and this process must be paid utmost importance to.

Knowledge must be very efficiently managed since this can be one of the major causes of the success/failure of an organisation. After all it is said that 'In Africa, when an old man dies, it is a library that burns down'.

7.12 Summary

KM plays a very important role in any organisation. It helps organisations retain and reuse the knowledge that they already have. This will help organisations in reducing cost and in increasing productivity and profitability. For effective and efficient utilisation of KM processes, KM implementation should be done in a systematic and holistic manner.

KM implementation process decides the success or failure of the KM processes. This document deals with KM implementation processes and frameworks in great detail. KM implementation process has ten golden steps which need to be followed for efficient utilisation of the KM processes. These steps have been covered in this document.

This document starts with a basic introduction to knowledge management. The document then covers KM frameworks for a corporate and a non-profit organisation. The most admired knowledge enterprise (MAKE) plan has also been mentioned. This is followed by the steps for implementation as mentioned above.

The document also gives a model to evaluate the implementation process. A roadmap for the KM implementation has been provided. This roadmap is just a basic reference, and each organisation will have to develop their own based on various factors such as resource availability, internal dynamics and organisational structure. The document also determines the probable gaps that can arise in the implementation process. These gaps must be avoided or rectified in the initial phases of implementation.

Most of the data in this document is based on classroom teaching, guest lectures and secondary research from the reference book and the Web resources.

Chapter 8
KM and Web 2.0

8.1 Web 2.0: The User-Driven Intelligent Web

The term Web 2.0 was first coined by O'Reilly's Dale Dougherty in a brainstorming session. It was basically a brainstorm session where this term was coined. Web 2.0 is much more than a just a new interface; rather it is a new way of thinking a new perspective on the entire business of software—from concept through delivery. Web 2.0 is databases that get richer the more people interact with them, applications that are smarter the more people use them, marketing that is driven by user stories and experiences and applications that interact with each other to form a broader computing platform.

It emphasises on sharing and collaboration and has its foundation on 'user-generated content'; it promises to promote a **bottom-up culture of knowledge sharing**. To state an example of what some experts have to say on the potential of Web 2.0, Don Tapscott, co-author of the popular management book *Wikinomics*, is of the view that 'the Facebook generation will wipe out the command control infrastructure in business today'.

The statistics of Facebook, Wiki, Myspace, etc. and other such sites also shows that given an opportunity the users are very proactive in sharing knowledge (Fig. 8.1).

8.2 Difference in Webs 1.0 and 2.0

The world is now talking of Web 2.0, but how is it different from Web 1.0 (though the term was never coined like this). While comparing we see that it is more of a shift from static to dynamic and also demands more user involvement (Fig. 8.2 and Table 8.1).

© Springer International Publishing Switzerland 2016
S. Mohapatra et al., *Designing Knowledge Management-Enabled Business Strategies*, Management for Professionals, DOI 10.1007/978-3-319-33894-1_8

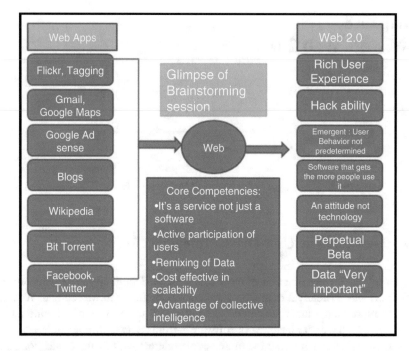

Fig. 8.1 Web 2.0

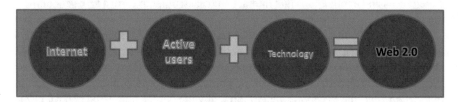

Fig. 8.2 Equation of Web 2.0

   Whereas Web 1.0 was static and read only, required knowledge of HTML was authoritative and had personal websites; Web 2.0 has brought **dynamism** in the whole World Wide Web.

8.3   Web 2.0 Deployment and Usage

After discussing about what Web 2.0 is and how it is different from Web 1.0, we wonder that for what purpose this can be used in the organisations. So we see that it can be used for content generation, community building and in decision support (Table 8.2).

Table 8.1 Difference between Web 1.0 and Web 2.0

Web 1.0		Web 2.0
Double click	➡	Google ad sense
Ofoto	➡	Flickr
Akamai	➡	BitTorrent
mp3.com	➡	Napster
Britannica online	➡	Wikipedia
Personal websites	➡	Blogging
Evite	➡	Upcoming.org and EVDB
Domain name speculation	➡	Search engine optimisation
Page views	➡	Cost per click
Screen scraping	➡	Web services
Publishing	➡	Participation
Content management systems (CMS)	➡	Wikis
Directories (taxonomy)	➡	Tagging ('folksonomy')
Stickiness	➡	Syndication

Table 8.2 Usage and possible uses of Web 2.0 deployment

Web 2.0 deployment	Usage	Possible uses
Content generation	*Mass internal communication*: Allow employees across the organisation to collaborate on codifying/ managing knowledge, sharing best practices, communicating and coordinating activities	Participatory knowledge management
	Tools: Broad collaboration/communication, metadata	
Community building	*Large-scale community building*: Enable large-scale, close-knit, distributed communities where experts are found	Lower attrition rates
	Tools: Broad collaboration/communication, metadata, social graphing	
Decision support	*Harnessing information markets*: Aggregate opinions from many individuals to guide strategic decision-making and idea generation	Faster decision-making and better return on investments
	Tools: Collective estimation	

8.4 Mashups: An Important Part of Web 2.0

Mashups combine data from multiple sites into single user experience. This user participation is reflected in the design of social networking sites: They operate as structured environments that allow users to manage their own Web pages—often with a number of configuration options for look and feel as well as various types of transaction processing. While there are many ways to capture and process information from other sites, the preferred Web 2.0 method is through a set of application platform interfaces (APIs) as one of the features of Web 2.0 is the ability to easily pull information from different sources

8.5 How API Is Different From Web

However, there is a significant architectural difference between traditional links such as those in online magazines that connect to ad servers and the way public APIs are being used by mashups: the actual retrieval of the information shifts from the client side to the server side. When a browser issues a request for a Web page, the browser interprets the content as it is downloaded. When it sees an external link such as those associated with advertisements, the browser issues a request directly to that site to retrieve the data.

In contrast, a mashup issues its request via a server-side 'proxy' or callback. The user still downloads the Web page, but in order to deal with 'same origin' requirements in JavaScript and other programming languages, the data retrieval must return to the original website. A proxied process in the background then retrieves the information and presents it back to the user. This means that several sources are trafficking data without explicit user direction.

8.6 Social Networking

After seeing some of the possible benefits like participatory knowledge management, lower attrition rates and better ROI, now we will discuss about social networking which is seen as a powerful form of Web 2.0 and knowledge management.

Social networking refers to the applications that are targeted to enable the creation and enlargement of a social network user. According to an economist survey, most of the multinationals have begun to see Web 2.0 technologies as corporate tools. Thirty-one percent of the respondents felt that using the Web as a platform for sharing and collaboration would affect all parts of their business.

8.6.1 Some Statistics

Internet 2.0

- Forty-two percent of office workers, aged between 18–29 years discuss work-related issues on social networking sites (blogs).

- Fifty percent of IT managers indicate that 30 % of bandwidth is social networking.
- Software as a service (SaaS) usage is steadily increasing.

Intranet 2.0

- Forrester predicts that Web 2.0 intranets will be a $4.6 billion industry within 5 years.

Extranet 2.0

- Nearly half of all Web developers are already using AJAX.
- In 2007 more than 30 % of large companies had a Web 2.0 business initiative.
- Sixty-six percent of companies indicate Web 2.0 is essential in maintaining their company's market position.

8.6.2 Social Networking Sites

- In the first quarter of 2006, MySpace.com signed up 280,000 new users each day and had the second most Internet traffic.
- By the second quarter of 2006, 50 million blogs were created—new ones were added at a rate of two per second.
- In 2005, eBay conducted eight billion API-based Web service transactions.
- There are 500 million plus users on Facebook, and so it has been called as the third largest country by population.

In September 2007, **Forrester Consulting** conducted a survey of 153 businesses with 1000 or more employees. They found that organisations, whether officially sanctioned or not, use Web 2.0 applications extensively. The future workplace will include Web 2.0-inspired applications such as RSS, blogs, RIAs, tagging, virtual worlds and wikis, according to a recent report by Forrester detailing the 'the seven tenets of the information workplace'.

8.7 Tools of Social Networking

8.7.1 Wikis

Wikis are revolutionising collaboration within the enterprise much as email has revolutionised communications. Wikis may be used for knowledge management, document management, project management, documentation, scheduling, meetings, directories and more. Research suggests that an estimated 33 % of enterprises are already using wikis and another 32 % plan to do so within 2 years.

8.7.2   RSS

RSS feeds are available for a wide variety of websites; many can be useful to professionals. In the workplace, professionals subscribe to a newsreader or aggregator client software or (if the feed is hosted) a Web application that aggregates syndicated. Web contents such as news headlines, blogs, podcasts and vlogs are pooled in a single location for easy viewing. Many companies that discovered utility in blogs and wikis are realising that RSS is necessary to push that content to users.

8.7.3   Tagging

Users and writers can tag every page read or written, using tags. Unlike the well-known world of taxonomy, where tagging is well defined by the organisation, tagging in Web 2.0 is rather personal. Everyone can tag (his or whoever's content), and the tags are also chosen personally and not from a predefined set of values. This collection of user-defined tags is called folksonomy.

8.7.4   Social Networking

All of the applications described above fit the definition of 'social networking' and all contribute to building this large net. Yet, the term, as known in the Web 2.0 world, refers to applications that are targeted to enable the creation and enlargement of the social networking (Table 8.3).

8.7.5   Example of Social Networking Websites

- LinkedIn.
- Facebook.
- MySpace.
- Twitter.
- Upcoming.
- Legal OnRamp ⎫ A network where lawyers interact.
- LawLink.
- Photo Sharing.
- Flickr.
- Podcasts: Stemming from the term 'broadcast', a podcast is a digital media file that is distributed over the Internet and listened to on a portable media player (like an iPod) or on a personal computer.
- YouTube.
- Xanga.

Table 8.3 Attributes and gaps of Web 2.0 components

Web 2.0 component	Attribute	Relevant attribute	Gaps
Wiki	Structured content pages	Web content management tools are a part of KM toolbox	Friendly user interface, flexible content and structure (also a disadvantage) and high level of interconnectivity
Blogs	Personal diary enables fast access to new pages and to history	Similar to the physical KM tool of storytelling	Very simple, accessible and appealing
RSS	Alerts regarding new content and changes by categories	Similar to alerts in enterprise content management tools as well as in portals	
Tagging	Everyone can tag his or her own information	Taxonomies are well defined, either at organisational or departmental level	Web 2.0 has no predefined list of attributes and values defining the 'allowed tags'
Social computing	Building social communities over the net	Communities of practice	Most Web 2.0 communities are based of hobbies or fields of interest. Communities of practice are more focused on professional issues, complemented by face-to-face gathering

8.8   Web 2.0 Technologies

See Table 8.4.

8.9   The Gap in Web 2.0 and Knowledge Management

The possible gaps that can be there are that knowledge is created but it is not validated or formatted for **reuse**. For that the companies have to make extra effort so that knowledge is shared, and also there are enough opportunities for people to create knowledge and **innovate**. It is very important that the knowledge that is created follows the whole process and is finally used; the knowledge created and not used will signify the gap between Web 2.0 and knowledge management. Say, for example, a blog, if the information is only created with no provision for its use, then this Web 2.0 technology will not become a part of knowledge management.

Web 2.0 is seen as a mechanism that elevates the practice of KM through better participation and usage by employees (Fig. 8.3).

Table 8.4 Web 2.0 technologies

Web 2.0 technologies	Description	Category of technology
Wikis, commenting, shared workspaces	Facilitates co creation of content/applications across large distributed set of people	Broad collaboration
Blogs, podcasts, video casts, peer to peer	Offers individuals a way to communicate/share information with broad set of other individuals	Broad collaboration
Prediction markets, information markets, polling	Harnesses the collective power of the community and generates a collectively derived answer	Collective estimation
Tagging, social bookmarking/filtering, user tracking, rating, RSS	Adds additional information to primary content to prioritise information or make it more valuable	Metadata creation
Social networking, network mapping	Leverages connections between people to offer new applications	Social graphing

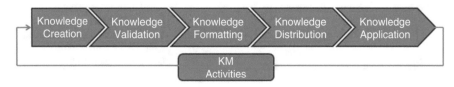

Fig. 8.3 KM activities

8.9.1   Verifying and Codifying Work: Some of the Best Practices that Can Be Integrated

- Do not allow anonymous contribution.
- Keep the power of policing in the hands of communities via flag as inappropriate option.
- Add an auditing for changes so that changes can be tracked.
- It is important that people understand the meaning of policy and basic monitor, and this should be cultivated in the organisation culture.

Researchers have pointed out that participation can be increased if Web 2.0 initiatives are aligned with the existing processes. In India there is a usage of blogs, wikis, discussion forums and shared workspace, but the use of RSS, social networking and podcasts has not developed much (Table 8.5).

Table 8.5 Best practices for blog and wiki

Best practices for blog	Best practices for wiki
• Code of conduct should be published of what can be blogged and what cannot be	• Provide auditing to rollback changes if necessary
• Provide anonymous moderators who can delete inappropriate content and comments	• Have an ability to override edit for selected topics or pages
• The blogs should have a searchable names and nothing abstract so that it becomes easy to search	

8.10 The Power of Network

8.10.1 Metcalfe's Law

Metcalfe's law states that the value of a telecommunications network is proportional to the square of the number of connected users of the system (n^2). He hypothesised that while the cost of the network grew linearly with the number of connections, the value was proportional to the square of the number of users. It was first formulated by George Gilder in 1993 and attributed to Robert Metcalfe in regard to Ethernet. It was originally given for Ethernet, but now it can be used with the emergence of Web 2.0. Metcalfe's Law can be used to explain the power and importance of all the communicating devices like fax machines; telephones; Internet platforms like Facebook, Orkut and Twitter; etc. The impact of knowledge management increases in direct proportion to the reach of knowledge flows through an organisation

Explanation of Power and Importance of Social Networking Sites on the Basis of Metcalfe's Law

Suppose there is only one person on Facebook and he can't communicate with anyone else, so it is useless; now if there are two, one connection can be made, and like this if the number of people on Facebook increases, the number of connections will increase like for five users and the number of connections which can be established is (5*4)/2 = 10; now when the number of users becomes 100, the number of connections which can be established is (100*99)/2 = 2475. This explains that when the number of users on the Internet increases, the number of connections which can be established is manyfold, so the information sharing becomes fast and plays a lot of roles in a person's life.

Table 8.6 IRR in social networking

Business challenge	Benefits of internal social networking	ROI
Change of staff, higher training cost	• More ideas are shared • More access to the knowledge experts	Lower attrition rates and therefore save training cost
	• Networking with the colleagues	
Conflicts, knowledge hoarding	• Can share ideas across all the divisions and branches	Good interpersonal relations in the organisation
Geographic and organisation divisions	• Information on shared platforms like blogs and wikis	Lesser hierarchical structures and greater encouragement to share knowledge and participate
	• Increase the company's knowledge pool	Employees more satisfied
Lost productivity and morale	• Consistent communication	Better productivity, profits, innovations
	• Perform job better	

8.10.2   Internal Rate of Return in Social Networking

Companies across all industries are experiencing the benefits of internal social networking; its interactivity connects employees with a breadth of valuable resources at a lower cost and faster speed. When used effectively, employees will share more ideas, make value-add connections, collaborate on projects and reduce the costs of communications, travel and operations.

Companies will retain a greater depth of knowledge, innovate faster, increase loyalty, reduce attrition, break down geographic and organisational divisions, improve relationships with customers and stakeholders and facilitate more effective ongoing communication.

Internal social networking helps companies create a connected workforce with unity of purpose to solve the business problems of yesterday and prepare for the opportunities of tomorrow (Table 8.6).

8.11   Limitation of Social Networking

Certainly one of the challenges of knowledge management is the ability to find and identify subject matter experts. These social networking sites are chock full of ways to find expertise.

In this part, social networking enables unprecedented power to explore a network that allows the exchange of ideas which can lead to new knowledge creation. However, this does not mean that when you find the right person, that person would actually be willing to help.

8.12 Communities of Practice

Certainly one of the challenges of knowledge management is the ability to find and identify subject matter experts. These social networking sites are chock full of ways to find expertise. Though Web 2.0 places a greater power in the hands of individuals; at the same time, it also increases the expectations from them.

The interface includes a community home page and **three additional tabs** (Fig. 8.4).

So we see that knowledge development in a COP is a continuous process and does not end at the organisation level. The individuals cannot only stay connected in the office but also join on Facebook, twitter, linkedin etc. so that the knowledge sharing can take place even when they leave the organisation.

8.13 Virtual Communities of Practice

Virtual CoPs may share news and advice of academic/professional interest but are unlikely to undertake joint projects together—this is more of the role of a distributed research centre. It may involve the conduct of original research, but it is more likely that its main purpose is to increase the knowledge of participants, via formal education or professional development. One critical success factor (CSF) for a virtual CoP is the technology and its usability. Other CSFs are after-action reviews and peer assists

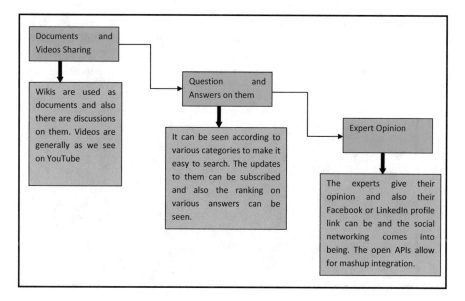

Fig. 8.4 Communities of practice

8.14   Enterprise Social Networking: Fluidity Versus Institutionalisation

Enterprise social networking tools on the other hand are aimed at connecting employees and emulating much of the functionality of online social networking sites. It aims at bringing the maximum gains for an organisation (Fig. 8.5).

Navigation the point of intersection between the two systems requires great leadership and management skill. The trade-off that an organisation also faces is the trade-off between fluidity and institutionalisation. The desired state would be the one which balances informal and formal network so that there is a good amount of bonding in the organisation and independence at any form to innovate but also people work constantly to achieve the business goals.

8.14.1   Examples of Organisation Providing Solution for Internal Social Networking

The social networking is very important in an organisation as we see that there are enterprises set up to, one such example is Kinetic Glue which will help every company set up its own 'official' equivalent of Facebook for employees to increase their productivity. There have also been some positive impacts as well.

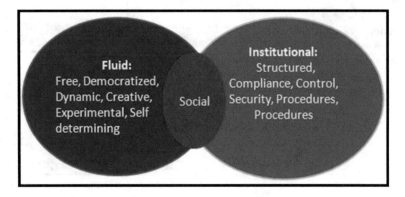

Fig. 8.5 Fluidity vs. institutionalisation

L&T Infotech which is an early user of Kinetic Glue quotes collaboration between 500 workers using this platform has helped it discover new revenue streams as it has enabled project groups within companies to have open conversations, with each discussion thread linked to relevant topics. Such system *breaks hierarchical barrier* as it encourages new entrants and less professional to contribute to the discussion. L&T also admits that their attrition has been lowered because of Kinetic Glue.

8.15   Benefits of Knowledge Management Through Social Networking

8.15.1   Six Degrees of Separation

The theory of "six degrees of separation" states that any two random-selected people on this world can get to know each other by no more than six steps of intermediate friend chains. Six degrees of separation, that is, that any two human beings, regardless of age, colour, creed or social status, have some sort of connection within five intermediaries. Through Web 2.0, this thing can very well be exploited to get better connections and hence better knowledge sharing.

8.15.2   Getting Knowledge on Time

It can improve the relations of people in a way that people are connected to more people and the people can be reached faster. It makes the knowledge sharing faster so that the knowledge reaches the people at the right time so that faster decisions can be made. An organisation generally has a lot of departments/divisions within it, and it faces problem in sharing the interdepartmental works. The problem can be addressed by social network in a way that the common blogs can be used for departments, and then people can know whom to contact and also reach faster. Better returns will be reaped if the social network is well designed.

8.15.3   Multitasking

The perfect time for many useful tools and Internet sites has enabled people to become more efficient at accomplishing multiple tasks in a shorter amount of time. The average worker spends nearly 2.5 h a day online (eMarketer)

8.15.4   Lowering Attrition Rate with the Help of Social Networking

As has rightly pointed out by L&T Infotech, many more have recognised that the attrition rate has been lowered because of the introduction of new social networking within the organisation. Not only does it remove the hierarchical structures but also encourage the new employees to contribute and when clubbed with the proper reward and recognition system will bring down not only the attrition rate but also retain the knowledge from going out.

8.15.5   Cost

The final theme was cost. That is, these social networking sites are free. So it is cheap and easy to experiment. You can see what things are useful to you and where the people you know are connecting. If the search options rather than metadata is not properly structured, then a company can lose up to $ six million per year in time spent just searching the information

8.15.6   Creating a Culture of Sharing in the Organisation

It helps in transitioning from an era of *knowledge is power* to *sharing knowledge* is more powerful and thus improves the way an individual thinks. As Web 2.0 focuses on removing the hierarchical barriers and encouraging participation from even the young entrants, it not only increases the social capital but also builds stronger bonds, and when people are given autonomy at work, they are not only motivated to do their work but they bring in innovation especially when the reward structure is organised properly.

8.15.7   Moving Beyond Geographical Boundaries

A shared space is created such that geographical boundaries do not create distance, and the knowledge flows without any such boundaries.

8.16   Importance of Data Management in Web 2.0

Web 2.0 is the term often used to describe the Web's evolution from a hypertext publishing system into a platform of participation and collaboration. In Web 2.0, the users are not only provided with the content but also contribute to the content by active participation. It is basically a two-way content transfer, and the content is no longer created locally and then published to the Web, but rather managed entirely online. In essence, Web 2.0, together with rich Internet applications (RIA) is taking over some of the functionality of traditional desktop applications. So the database management becomes a core competency of Web 2.0 companies. These companies made substantial investments in their databases. NavTeq alone reportedly invested $750 million to build their database of street addresses and directions. Digital Globe spent $500 million to launch their own satellite to improve on government-supplied imagery, and similar is the story with Facebook which not only has to manage the data about the individuals but all the photos, videos, comments, games, advertisement, etc. that is put in that which is in itself a huge process.

8.16.1   Managing of Data by Social Networking Sites

In particular, several Web 2.0 sites have emerged that allow personal information to be managed collaboratively over the Web. Personal data can be of different types and can be categorised into two main groups.

First, there is data which formerly has been managed locally by desktop applications, such as contacts, documents or pictures. Second, there is a lot of personal data which has only emerged because of the Web, such as bookmarks or any kind of metadata associated with media content.

Web 2.0 applications that manage such personal information include sites such as Flickr and YouTube for images and videos, Blogs for writing diaries or travel journals, Google Documents for managing and sharing documents or del.icio.us for Web bookmarks. It is important to note that while some of these sites focus on a user's private life, others are closely related to work activities. Thus, friendships might be managed by Facebook, while professional networks and contact information may be managed by sites such as LinkedIn or Xing. Clearly, this development in terms of how the Web is used for personal information management (PIM) also raises several technical challenges with regard to data management.

While the Web is an ideal platform for collaboration and participation, Web applications generally cannot compete with desktop applications in terms of complexity and integration into the local working environment. Further, while a few approaches have been proposed to enable users to work with Web 2.0 applications offline, currently desktop applications still cope better with this requirement.

Therefore, it is not to be expected that Web 2.0 applications will replace desktop applications entirely, but rather that the two kinds of applications will be used for different tasks and modes of working. However, they may share data which means that, ideally, there should be some way of seamlessly managing data across desktop applications and Web 2.0 applications. The distribution of information across desktop applications and Web 2.0 applications introduces further forms of information fragmentation already considered to be one of the main issues of PIM with respect to desktop applications.

8.16.2   New Forms of Data Management Structures Required

We believe that new forms of data management architectures are required that can provide an integrated approach to data management for desktop and Web 2.0 applications with a clear separation of concerns between the management of personal data and its publication on the Web to social networks. One of the main technical requirements is to provide ways in which data can be synchronised between desktop applications and Web 2.0 applications. However, such an approach should also incorporate concepts that have proven useful in Web 2.0 applications. One such feature that can be witnessed on sites such as Facebook is the ability to extend these platforms based on plug-ins or modules. In order to successfully manage data for Web 2.0, we believe that the data management architecture should reflect this notion of components.

8.17   Limitations of Web 2.0 in Knowledge Management

Many organisations want benefits of Web 2.0 but are worried about the impact. Web 2.0 has invaded the workplace leading to a dramatic increase in malware and data leak vulnerabilities; Web 2.0 sites often use data from different sources, such as a retail store site using Google maps to display locations; this makes it more difficult for security systems to validate the integrity of the code. Hackers are also embedding spam and malicious code into other types of content, such as instant messaging, shared video content and business documents like PDF and Excel files; criminal hackers have cost United States businesses an estimated $67.2 billion a year, according to the US Federal Bureau of Investigation (FBI).

8.17.1   Bandwidth Utilisation

Web usage and bandwidth consumption in turn lead to decreased employee productivity. In a survey of network managers and security managers, more than 70 % wanted social networking sites banned (Secure IT).

8.17.2   Loss of Productivity

Numerous Web 2.0 applications are useful for business purposes. At the same time, however, the extensive use of Web 2.0 applications leads to increased non-business Web usage and bandwidth consumption, which in turn leads to decreased employee productivity. Many businesses believe that taking away access to social networking and rich media sites will be visible.

8.17.3 More Malware Access

Web 2.0 includes, at minimum, the technologies associated with client-side processing, syndication, mashups and user participation:

- Client-side processing—more surface areas for attack
- Syndication—greater transparency for attackers
- Mashups—complex trust scenarios
- User participation—erosion of traditional boundaries
- More avenues for data leakage—wikis, blogs, trackbacks (described further in this document), emails, instant messaging

8.17.4 Too Many Passive Employees

Web 2.0 requires participation from people, and if there are a lot of passive employees whether managers or employees who understand the importance of knowledge management but do not have time or attentiveness to join in, then the purpose of Web 2.0 is not resolved.

8.18 Summary

Web 2.0 has taken social networking by storm. The concept has created virtual forum that has connected people for different causes. It's natural that organisations take cognisance of this technology to reap business benefits. One of the areas for early adaption of Web 2.0 can be knowledge management area, where employees would connect through Web 2.0 in virtual space. They will create community of practice in a virtual manner to share and learn from each other. However, to align the strategy with virtual technology, a systematic approach is required. This chapter explains the step by step methodology that can be used for using Web 2.0 for knowledge management.

Chapter 9
KM and Cloud Computing

9.1 Introduction

In June 2009, a study conducted by Version One found that 41 % of senior IT professionals actually don't know what cloud computing is and two-thirds of senior finance professionals are confused by the concept, highlighting the young nature of the technology. The technology though remains unknown to most of us is definitely something that must not be and cannot be ignored.

Cloud computing may mean different things to different people, but we all must agree at this that it has opened up new avenues for individuals as well as organisations offering them flexibility, cost efficiency and unimaginable opportunities. In Sept 2009, an Aberdeen Group study found that disciplined companies achieved on average an 18 % reduction in their IT budget from cloud computing and a 16 % reduction in data centre power costs. Online office suites, such as Zoho and Google Docs, and even online operating systems such as eyeOS have been able to successfully provide for the needs of varied customers.

Cloud computing can be defined as technology that uses the Internet and central remote servers to maintain data and applications, thus allowing consumers and businesses to use applications without installation and access their personal files at any computer with Internet access. This technology allows for much more efficient computing and reduced cost by centralising storage, memory, processing and bandwidth. It encompasses many areas of tech, including software as a service, a software distribution method pioneered by Salesforce.com about a decade ago. It also includes newer avenues such as hardware as a service, away order storage and server capacity on demand from Amazon and others. What all these cloud computing services have in common, though, is that they're all delivered over the Internet, on demand, from massive data centres. The software and the services are provided on a 'rental' basis via the Web.

The term cloud computing was inspired by the cloud symbol that is often used to represent the Internet in flowcharts and diagrams. Cloud computing can be conceptually represented by the following diagram (Fig. 9.1).

© Springer International Publishing Switzerland 2016
S. Mohapatra et al., *Designing Knowledge Management-Enabled Business Strategies*, Management for Professionals, DOI 10.1007/978-3-319-33894-1_9

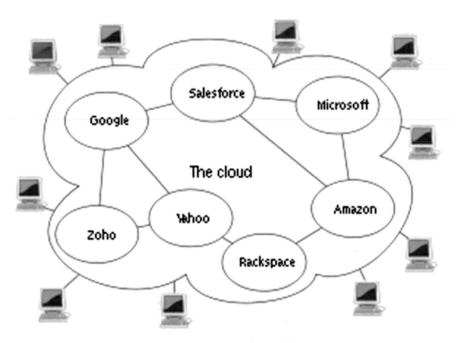

Fig. 9.1 Cloud computing conceptual diagram

In general, cloud computing customer avoids capital expenditure of the company, thereby also reducing the cost of purchasing physical infrastructure by renting the usage from a third-party provider. The companies devour the resources and pay for what they use. Many cloud computing offerings use the utility computing model, which is similar to traditional used services like electricity that are consumed, while others bill on a subscription basis. So sharing 'unpreserved and intangible' computing power among multiple customers can improve operational rates. A side effect of this approach is that the overall computer usage has risen severely because the customers do not have to engineer for peak load limits.

A layman example for cloud computing is our sending of emails on Gmail, Yahoo!, etc. The server and email management software is all on the cloud and managed by cloud service providers such as Yahoo! or Gmail. This does away with the need of having one's own storage space and gives the convenience of sending and receiving mail at minimal cost.

9.2   Data Centre

The original data centre was an organisation's private facility consisting of several servers performing the role of maintaining and running dedicated functions each. This took up lots of resources and personnel of organisation which had the herculean task of maintaining and running the servers as well as the facility. While such data centres continue to exist in some larger organisations, many others have been able to increase service levels, reduce the response time and increase the number of users covered by outsourcing the job to third-party data centres and cloud computing providers. These third-party data centres can cater to the organisation's requirements as they are better equipped to maintain and update server equipment.

A data centre also referred to as a server farm is a centralised repository for the storage, management and dissemination of data and information. Usually, it is a facility used to contain computer systems and associated components such as telecommunications and storage systems. Principally, it is an aggregation of physical hard drive storage resources into storage pools from which logical storage is formed. To ensure a steady and quality service, there are redundant or backup power supplies, redundant data communications connections, environmental controls and security devices.

Many different vendors can be served by a single data centre as addition of vendors' storage hardware to the system has inconspicuous effect. Many different computer systems that share the same pool of storage space can reach these logical storage spaces by their respective systems. Virtualisation, hence, offers benefits such as centralised backups and reduction in a number of hard drives. Replication and migration of data to another location transparently to the server can be done using the logical storage point. The benefits of the data centre can be attributed to the fact that it is the consolidation of all of the facility resources such as HVAC, electrical, hardware, software, network connections, wiring and personnel. Many organisations have multiple server rooms with duplicated services across their entire organisation, all of which are running on duplicated hardware and software platforms. Many corporations are consolidating their server rooms into private data centres, reducing the duplication of hardware, software and facilities needed to operate their business. This in turn results in reduction in expenses and freeing of resources which can be deployed in the core business.

9.3   Cloud Computing

The concept of cloud computing involves a data centre somewhere in the world or even multiple data centres scattered around the world. The data centres as well as the clients form a part of the cloud. The historical client-server architecture where the network users owned, maintained and operated their own network infrastructure, server rooms and data servers and applications were expensive to maintain and run and resources intensive. Leveraging third-party computing facility and expertise over the network is a good way to cut costs, increase scale and improve responsiveness.

Typical cloud computing providers deliver common business applications which can be broadly divided into the following categories: software as a service (SaaS), utility computing, Web services, platform as a service (PaaS), managed service providers (MSP), service commerce and Internet integration. These applications are delivered online and are accessed from Web browsers, while the software and data are stored on the servers.

Cloud Computing

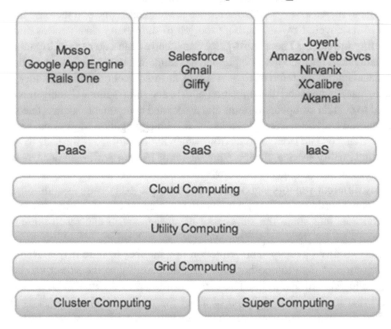

Fig. 9.2 Cloud computing

These data centres host the servers and applications which are utilised by the clients in their business. This structure is a lucrative option for the corporations as it reduces their capital expenditures while also providing more responsiveness. It is a win situation as the third party earns by charging rental, whereas the organisation earns by paying only for resources used. While some cloud providers employ a utility computing model for billing the users, others bill on a subscription basis. By paying rental to the provider, the customer gains the security of a service level agreement (SLA) as well as the saved expense of personnel and resources (Fig. 9.2).

A client has the flexibility to purchase or rent several resources available in data centre, such as processing time, network bandwidth, disk storage and memory. The users of the cloud only need to know how to connect to the resources and how to use the applications needed to perform their jobs. The location of the data centre or the know how to operate or maintain the resources in the cloud, is not required by him. This gives convenience of simplicity to the clients.

In a cloud, the user's computer runs a user interface. It provides a window into the application, but does not actually run the application. The applications are actually run on servers in the data centre, which are often shared by many users. This procedure implies centralisation of data centre and reduces the hardware requirement such as big processing power and memory on the end user's computer.

9.4   Virtualisation

Virtualisation is one of the main cost-saving, hardware-reducing and energy-saving techniques used by cloud providers. Virtualisation is carried out with software-based computers that share the underlying physical machine resources among different virtual machines.

With operating system virtualisation, usage of different operating systems and isolation of each operating system is possible. Many companies enable different services to run in separate virtual machines on the same physical machine by using virtual machines to consolidate servers. Virtual machines also provide time sharing of a single computer among several single-tasking operating systems. Usage of virtual machines requires the guest operating systems to use memory virtualisation to share the memory of the one physical host.

Memory virtualisation creates virtualised memory pool by removing volatile random access memory (RAM) resources from individual systems and aggregating those in the pool. This virtualised memory pool is available to any computer in the cluster. Memory virtualisation thus, by leveraging large amount of memory, improves overall performance, system utilisation and efficiency. By avoiding replication of shared data of multiple servers also reduces the total memory requirement.

9.5   Types of Clouds

There are three types of clouds which are deployed (Fig. 9.3):

- Public
- Private
- Hybrid

9.6   Public Cloud

Public cloud or *external cloud* describes cloud computing in the traditional mainstream sense, whereby resources are dynamically provisioned on a fine-grained, self-service basis over the Internet, via Web applications/Web services, from an off-site third-party provider who bills on a fine-grained utility computing basis. These clouds offer the greatest level of efficiency in shared resources; however, they are also more vulnerable and are often held up for not providing privacy and security. A public cloud is deployed when standardised workload for applications is used by lots of people, such as email. It does not need customisation and is useful for a large number of clients. It is generally used to provide incremental capacity so that the organisation can go on without interruption even when it has reached its computer capacity. Also it is preferred when the organisation is having a

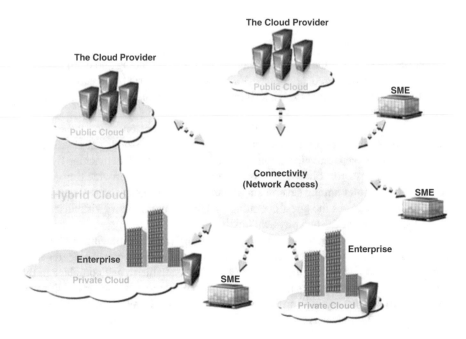

Fig. 9.3 Types of cloud

collaboration project or when it needs to test and develop application code. It is advisable to use public clouds if the company already has SaaS from a vendor who has a well-implemented security strategy.

Many organisations for which it is important to maintain information confidentiality are concerned about public cloud security and reliability. It must be always ensured that one has security and governance issues well planned to avoid any business catastrophe.

9.7 Hybrid Cloud

The term 'hybrid cloud' refers to the use of physical hardware and virtualised cloud server instances together to provide a single common service. It includes a variety of public and private options with multiple providers. Hybrid storage clouds are often useful for archiving and backup functions, allowing local data to be replicated to a public cloud.

A hybrid cloud can ensure security for an organisation that wishes to use a SaaS application. The SaaS vendor can create a private cloud for the company inside their firewall and provide it with a virtual private network (VPN) for additional security. A hybrid cloud can provide a company with the environment that it needs to better serve its customers. While offering services that are tailored for different vertical markets, the organisation can decide to use a public cloud to interact with the clients but keep their data secured within a private cloud.

By spreading things out over a hybrid cloud, one can keep each aspect at business in the most efficient environment possible. The downside, however, is that one has to keep track of multiple different security platforms and ensure that all aspects of business can communicate with each other. The management requirements of cloud computing become much more complex when one needs to manage private, public and traditional data centres all together. There arises a need to add capabilities for federating these environments.

9.8 Private Cloud

A private cloud is one in which the services and infrastructure are maintained on a private network. These clouds offer the greatest level of security and control, but they require the company to still purchase and maintain all the software and infrastructure, which in turn increases the cost. Thus it must be deployed by a company that is large enough to run a next-generation cloud data centre efficiently and effectively on its own.

A private cloud is the obvious choice when data and applications form the most important part of business thereby making control and security of paramount importance. An organisation which is a part of an industry that must conform to strict security and data privacy issues has no other choice but to go for private cloud.

Private clouds which do not benefit from lower upfront capital costs and less hands-on management actually lack the economic model that makes cloud computing such a fascinating concept.

9.9 Segments of Cloud Computing

Cloud computing typically consists of three segments (Fig. 9.4).

Applications—It is based on 'on-demand' software services. On-demand software services come in different varieties. They vary in their pricing scheme and how the software is delivered to the end users. In the past, the end user would purchase a server that can be accessed by the end user over the Internet. Salesforce.com (CRM), Google (GOOG) and NetSuite (N) are some of the companies which provide on-demand software or SaaS business.

Some of the companies which are already established in the on-demand software or SaaS business include Salesforce.com (CRM), Google (GOOG), NetSuite (N), Cordys, Taleo (TLEO) and Concur Technologies (CNQR). These companies charge their customers a subscription fee and in return host software on central servers that are accessed by the end user via the Internet.

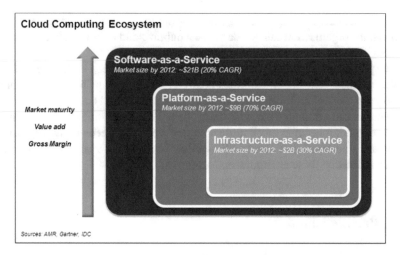

Fig. 9.4 Segments of cloud computing

Some of the companies who have established themselves as traditional software providers include SAP AG (SAP), Oracle (ORCL), Blackbaud (BLKB), Lawson Software (LWSN) and Blackboard (BBBB). These companies sell licences to their users, who then run the software from on-premise servers.

Platform—Platform as a service (PaaS) is the service and management layer of the cloud platform and is evolving dynamically to include things such as intelligent provisioning, as well as application and network management. It refers to products that are used to deploy the Internet. Many of the companies that started out providing on-demand application services have developed platform services as well. NetSuite, Amazon, Google and Microsoft have also developed platforms that allow users to access applications from centralised servers.

Active platforms—The following companies are some that have developed platforms that allow end users to access applications from centralised servers using the Internet. Next to each company is the name of their platform:

- Google (GOOG)—Apps Engine
- Amazon.com (AMZN)—EC2
- Microsoft (MSFT)—Windows Azure
- SAVVIS (SVVS)—Symphony VPDC
- Terre mark Worldwide (TMRK)—The Enterprise Cloud
- Salesforce.com (CRM)—Force.com
- NetSuite (N)—SuiteFlex
- Rackspace Cloud—cloudservers, cloudsites, cloudfiles
- Metrisoft—Metrisoft SaaS Platform
- SUN Oracle direct link
- Cordys Process Factory—The Enterprise Cloud Platform

Infrastructure—The third segment in cloud computing, known as the infrastructure, is the backbone of the entire concept. Infrastructure as a service is the foundational layer of cloud computing and includes raw storage, compute, backup, disaster recovery, databases and security infrastructure vendors' environments such as Google gears that allow users to build applications. Cloud storage, such as Amazon's S3, is also considered to be part of the infrastructure segment.

Major Infrastructure Vendors—Below are companies that provide infrastructure services:

- Google (GOOG)—Managed hosting, development environment
- International Business Machines (IBM)—Managed hosting
- SAVVIS (SVVS)—Managed hosting and cloud computing
- Terremark Worldwide (TMRK)—Managed hosting
- Amazon.com (AMZN)—Cloud storage
- Rackspace Hosting (RAX)—Managed hosting and cloud computing

As the first segment to emerge in scale and the most application oriented, SaaS has led the market to date with the largest market size, highest gross margins, and highest per-seat pricing. Recently, however, we've seen the rapid emergence of hyper-growth businesses in the PaaS and SaaS markets demonstrating that these will soon be independent, multi-billion dollar segments in their own rights with the potential for massive sales volume and attractive cash flow characteristics.

9.10   KM and Cloud Computing

Knowledge management comprises a range of strategies and practices used within an organisation to identify, create, represent, distribute and enable adoption of insights and experiences. Such insights and experiences comprise knowledge, either embodied in individuals or embedded in organisational processes or practice (Fig. 9.5).

Fig. 9.5 KM and cloud computing

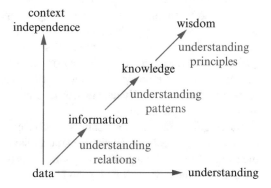

By implementation of knowledge management, an organisation can successfully utilise new ways to channel raw data into meaningful information. It is this information that can then become the knowledge that leads to wisdom. In regard to accessing the worth of knowledge, one must look it through the system thinking. The point is that information, knowledge and wisdom are more than simply collections. The whole represents more than the sum of its parts. While information relates to description, definition or perspective and addresses the questions of what, who, when and where, knowledge comprises strategy, practice, method or approach, thereby answering the question of how. Wisdom, eventually, embodies principle, insight, moral or archetype and gives vision to answer the why.

Knowledge management enhances an organisation's ability and capacity to deal with and accomplish the critical parameters of organisational success such as:

- Mission: What are we trying to accomplish?
- Competition: How do we gain a competitive edge?
- Performance: How do we deliver the results?
- Change: How do we cope with change?

The effectiveness with which the managed knowledge enables the members of the organisation to deal with present situations and effectively envision and create their future is the parameter to judge efficacy of knowledge management. In the absence of on-demand access to managed knowledge, every situation is addressed based on what the individual or group brings to the situation with them. With on-demand access to managed knowledge, every situation is addressed with the sum total of everything anyone in the organisation has ever learnt about a situation of a similar nature. This ensures timely availability of suitable information at the time of need. The very nature of today's 'on-demand' environment is that it is subject to continuous change. In the past, the task of knowledge management that could suitably be properly utilised by end users and business executives to meet their day-to-day challenges was a herculean effort that produced limited results. Today, IT offers solution to the management systems which must oversee a combination of static systems and ad hoc resources, ranging from virtualised applications to third-party infrastructure. Effective IT management systems must be easier to administer and more responsive and quicker to produce timely analysis. Today's cloud computing movement offers exciting opportunities to meet these age-old challenges.

With economic uncertainties, escalating competition, declining customer loyalties and an increasingly dispersed workforce, the pressure to successfully address the issue of knowledge management has never been more severe. Adding to this challenge is the limited success many organisations have achieved in implementation of knowledge management to guide their day-to-day activities and long-range initiatives. Technology, finance as well as employee behaviour all threw huge challenges to the implementer.

Moving KM into the cloud should be a business decision rather than an information technology decision. Businesses must take a more strategic approach for what the cloud can do for our business opportunities, just like what they did with knowledge management. It can't be just an IT thing.

It is communities of practice or interest which is the nexus between cloud computing and knowledge management. One solution to these challenges has been in the form of social networks. Sites and services such as Facebook, Twitter and LinkedIn have made information sharing second nature in our personal and professional lives. However, the use of these online services as business tools has typically evolved outside the realm of conventional enterprise applications. As a result, many organisations are still grappling with how to integrate the information and harness the insight being generated by these services into their corporate operations. They are also trying to develop the appropriate policies and procedures to protect their proprietary interests and adhere to the compliance requirements within their industries. Today's cloud computing alternatives are beginning to untangle these issues and seem to be promising.

9.11 Benefits of Cloud Computing

Cloud computing is already becoming a reality for large enterprises as it promises benefits such as reduced costs, on-demand availability of services, use of services and infrastructures as and when required and re-allocation of costs from capital investment to operational expenditure. By the virtue of cloud computing, an organisation can reduce its cost of operation and need for technical staff and increase the amount of time and energy expended on their core business.

The use of cloud computing also has many other benefits, including increased productivity, since users can access their applications from anywhere on the Internet. It reduces the need for hardware by 'time sharing' clients on the same hardware platform with the use of virtualisation. By reducing the number of servers, the data centre reduces its need for power and HVAC, again reducing the cost of doing business and also doing its share to save the planet.

For those whom frequent travelling is part of the job description, accessing much-needed tools and data on the field is a big challenge. External devices can only hold so much, and the threat of theft is always present. It may not be worth the risk if one is dealing with sensitive information. Cloud computing and cloud storage make such issues manageable.

Today, consumers are accumulating massive amounts of personal data, and these are usually distributed among several computers, digital media players, mobile phones, home entertainment centre and so on. Since these devices tend to overlap on certain tasks, data duplication is the norm. For example, music can be played on an iPod and on a MacBook. Documents can be edited on a PC and on a smartphone (Fig. 9.6).

Key benefits of cloud computing can be stated as:

- Agility—This is the ability to update hardware and software quickly to adhere to customer demands and updates in technology. This has provided immense competitive edge to the organisations.

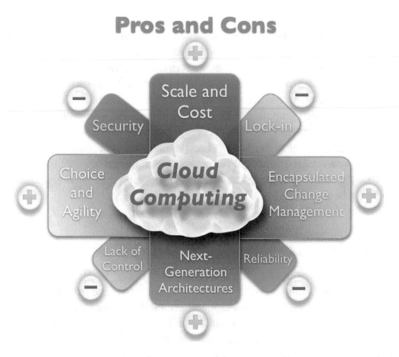

Fig. 9.6 Pros and cons of cloud computing

- Savings—Due to shared resources, a reduction in requirements of capital expenditures and IT personnel occurs. This capital as well as personnel can be deployed for core business of the organisation.
- Device and Location Independence—Users can access application from a Web browser connected anywhere on the Internet. This has increased the freedom to move around without bothering about the hardware constraints.
- Multi-tenancy—Resources and cost are shared among many users, allowing overall cost reduction. A user pays only for what he uses. This makes cloud computing a much resorted option.
- Reliability—Many cloud providers replicate their server environments in multiple data centres around the globe, which ensures that the services do not get disrupted even in case of disasters.
- Scalability—Multiple resources load balance peak load capacity and utilisation across multiple hardware platforms in different locations gives the organisation scalability to the extent desired by it.
- Security—Centralisation of sensitive data improves security by removing data from the users' computers. It has been known that most of the confidential data are leaked through the organisation personnel. By protecting the data on centralised servers, cloud computing ensures that data remains secure. Cloud providers also have the staff resources to maintain all the latest security features to help protect data.

- Maintenance—Cloud computing involves centralisation of applications. As it is much easier to maintain applications centrally than their distributed counter parts, easy maintenance further increases the attractiveness of the technology. All updates and changes need to be made in one centralised server instead of on each user's computer.

However, cloud computing must not be only seen as a bundle of opportunities. It can prove problematic and disastrous if it is not handled carefully. Lack of control on data can give rise to various security issues. Thus, the opportunities offered by cloud computing come with problems too and require regulation and monitoring to ensure that data is not misused.

9.12   Introduction of Cloud Computing in Indian IT Firms

With its advent, subsequent reputation and adoption, cloud computing has changed the role IT plays in a firm and the way firms should handle it going forward. IT, which in most cases has been a vertical supporting the core business of a company (even in a lot of technology firms), is now triggering changes in the business model of an organisation. Cloud computing is helping firms stay cost-effective, efficient and adaptable to changes by redesigning their IT model. In this short paper, we would present what changes cloud computing has brought, how it is changing the role of existing technologist in a firm and what the future holds.

Ever since its advent, cloud computing has gained rapid popularity and has helped a lot of firms reduce operating costs, enhance profitability, smoothen growth, increase efficiency and reduce operational and managerial complexity. For existing firms, this requires moving to an adapted business model, in which they can align their business goals with the IT infrastructure from the cloud, resulting in increased efficiency and reduced costs. For start-ups, cloud computing has been all the more important. Cloud computing very well handles the high capital investment risk problem which a lot of companies face during their inception and initial years. Through its pay as you use model, cloud computing takes care of this uncertainty and helps firms become less risk averse. Moreover, for a start-up, it is easier to design its business model which is cloud ready than for a firm which is already operating at a large scale. This reasoning is also reaffirmed by the fact that a large number of start-ups now prefer to use the cloud as their IT infrastructure system (Fig. 9.7).

9.13   Redefining Role of IT Within a Firm

For long, IT has been a support vertical in a firm. Firms develop their core business model and then use IT as a support shaft for executing their plan. Decisions of what kind of IT infrastructure to use and how to deploy them have been an important decision but primarily with the focus of making sure that the fleet of machines the firm

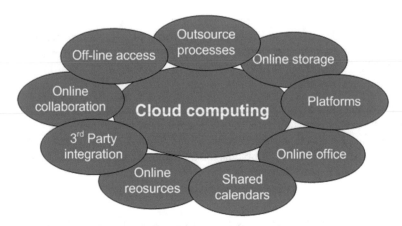

Fig. 9.7   Features of cloud computing

owns don't age out around the same time. Secondly, firms also maintain redundancy in order to ensure high availability and quality of service. Ensuring both of these meant higher investment and higher cost of maintenance for the firm. Cloud computing is primarily targeted towards solving this very problem. Through its adoption, it changes the role of IT in a firm. Once on a cloud, firms can scale up or scale down their IT fleet very quickly and hence becomes highly adaptable to changes. Many IT organisations are moving their IT infrastructure to a private cloud (closed and shunned from the Internet). This has helped them address some of the problems mentioned above. Companies are availing high availability and reliability at a lesser cost by significantly reducing their cost on building IT infrastructure and its subsequent maintenance. Moving their infrastructure to the cloud gives them the advantage to expand or contract their infrastructure on a need basis for, e.g. expansion in case of a rapid growth in employee base. It is an absolute must for companies whose IT resources, computing and storage needs are very unpredictable and vary around the year. For organisations with a relatively constant utilisation of IT infrastructure, the benefits would be in terms of optimisation of its resources. Patni Computer Systems has reduced its CapEx spending by 30 % and lowered its power consumption by 30 %. Cloud-based infrastructure will also help them during mergers and acquisitions to simplify the data aggregation. Under this changed schema of things, the role of a technology officer or a CTO becomes even more crucial. Under the older model, firms used to plan their IT resources for the next 6 months or 1 year and usually stacked a significant over capacity of resources. Now the need is to keep constant tab on the scale of business activity and change IT resources together with co-related change in the business. This has considerably reduced the reaction window and also the requirement of technology officer to work more closely with business development leader.

Another argument here can be that now there are lots of cloud brokers, which help firms in the migration to cloud from its historical data centre-based model. This may very well make the role of a technology officer redundant. Though the argument looks strong at the face of it, we still doubt it because not many firms would

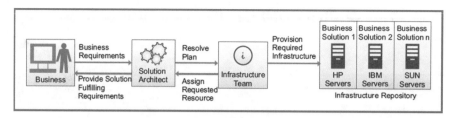

Fig. 9.8 Infrastructure provisioning: traditional model

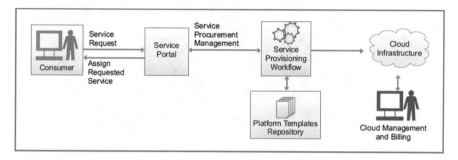

Fig. 9.9 Infrastructure provisioning: cloud computing

leave crucial decision of managing their IT fleet and resources on to cloud brokers. They may end-up using cloud brokers together with human management, which still significantly increases the importance of a technology officer in the firm (Figs. 9.8 and 9.9).

9.14   Role of IT Vendors in Cloud Computing

IT solution providers have started offering cloud-based infrastructure management services. Most providers offer infrastructure management services for private clouds hosted on client's own infrastructure as well as for private clouds hosted on the service provider's network (referred to as infrastructure as a service).

Big IT players such as Microsoft, Google and Amazon that have abundant resources in their network are offering platform as a service and infrastructure as a service. These clouds are also referred to as public cloud. Some of these services are free, and some are based on pay-per-use model (Fig. 9.10).

Outsourcing solutions based on cloud have also become very common. Many vendors such as TCS, Wipro, and Patni Computer Systems offer solutions in this space. There are a wide variety of offerings available based on fixed cost, variable cost (pay-per-user, pay on time basis, pay on bandwidth usage, etc.) and a combination of the two. These offerings are mostly in the bracket of software as a service. As an example on the revenue potential for IT players in cloud computing, TCS has

THE CLOUD COMPUTING ADOPTION MODEL

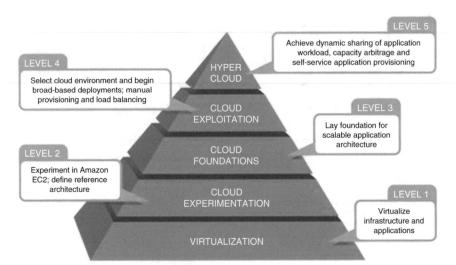

Fig. 9.10 Cloud computing adoption model

forecasted 10 % of its incremental revenue from cloud solutions in the next year. This shows the present revenue and future potential of cloud computing for Tata Consultancy Services, a major Indian IT service player.

9.15   Cloud Computing Can Close the Development Gap

Despite tremendous global economic growth over the last few decades, prosperity has evaded huge swaths of the world. Hundreds of millions of people still live in abject poverty with limited prospects for social and economic development. It was in recognition of this imbalance that the United Nations adopted the Millennium Development Goals (MDGs) at the UN Millennium Summit in 2000. However, more than midway through the 15-year timeframe set to realise the MDGs, progress against the goals—whether reducing extreme poverty, providing universal primary education or advancing the quality of healthcare—continues to be uneven at best. The substantial gap between aspiration and reality was underscored at the recent session of the United Nations General Assembly in New York in September 2008. The current financial meltdown and the worldwide economic recession will only exacerbate this situation, as there will be even fewer resources available to close the development gap. 'Cloud computing'—a paradigm shift now occurring in the information technology (IT) industry—offers the strong possibility of accelerating social and economic development, even in this time of limited resources. Historically, development agencies and nongovernmental organisations (NGOs), especially in developing countries, have not been able to fully leverage information technology to further their goals and objectives, primarily owing to the significant cost and complexities in deploying and managing IT. Now, as the information technology industry goes through a major shift, founded on the Internet as a platform, new opportunities are open for them to employ technology at a lower cost and with much greater ease and success than in the past. While ubiquitous and affordable Internet access—as well as reliable electricity—is not yet a reality today, there are sufficient pockets of the developing world that are equipped to take advantage of this new approach to delivering and consuming IT. Governments around the world should actively promote policies and partner with the private sector to accelerate the availability of Internet access to all citizens and ensure that the new information technology model known as cloud computing is not hampered—intentionally or unintentionally—by the evolving regulatory environment around the Internet.

9.16   Cloud Computing for Development

The traditional 'ownership' model of technology has presented significant obstacles for development agencies and NGOs to consistently and broadly exploit IT. Not only are the barriers to entry very high, but the resources required over time to sustain such technology deployments are often prohibitive. The technology industry has played a critical role in trying to alleviate some of these issues. Companies such as Cisco, HP, IBM, Intel, Microsoft and others have a strong record of supporting social and economic development initiatives around the world as part of their corporate philanthropic efforts. They have made substantial contributions over the years in the form of donations of equipment and infrastructure, as well as strategic engagement on development projects to offer direction, advice and expertise. Nevertheless, the complexity of IT deployments, the expertise required to maintain IT systems over the course of their lives and the resources required to support IT users have often rendered the use of information technology outside the reach of development organisations and prevented IT from being used pervasively, particularly in developing countries.

As with consumers and businesses, cloud computing holds tremendous promise for development organisations, including agencies and NGOs in developing countries. Development organisations already use the Internet in various ways. In the early days of the Internet, for example, many development agencies and NGOs were quick to set up websites to broadcast their message and reach out to a wider audience. Today, of course, having a website is as common as printing a brochure, and development organisations use websites to provide detailed information on their activities and programmes. The Internet is also increasingly being used as a means to engage in trading and commerce to advance the economic prospects of rural communities. A case in point is an initiative in India called e-Choupal, which provides vital information on crop prices, weather conditions and scientific farming practices to 3.5 million farmers across 31,000 villages and allows them to use an e-trading service to get the best prices in selling their crops over the Internet. More recently, the Internet has been used as a communication channel to deliver higher-quality social services to people in rural areas. Telemedicine, which allows people to connect over the Internet to receive medical advice from specialists thousands of miles away, is one such example. Although the Internet is currently being used in these ways to further social and economic development, such efforts are still largely rooted in the 'ownership' model of information technology—albeit with the Internet as the communication network to transfer data. Cloud computing now enables development organisations to greatly expand their capabilities by deploying sophisticated information technology solutions without the cost and complexities of purchasing and setting up IT systems. In the same way that companies are now using Web-based services to manage their business operations, development organisations have the opportunity to improve the efficiency of their internal operations using software solutions offered as a service over the Internet. In addition to using pre-built software over the Internet, development agencies and NGOs can use online services to build customised Web-based software programmes for their own specialised needs—ranging from fund-raising and grant management to volunteer programmes and project management—with little coding or technical resources, similar to the way individuals create personalised Web pages on Google and Yahoo! with point-and-click ease. Using software in this manner allows development organisations with limited resources to benefit from regular software updates and innovations without additional expense or disruption to operations.

Beyond improving internal operations, cloud computing can be employed to promote development initiatives and achieve higher levels of social and economic progress in disadvantaged communities. Just as mobile phones enabled communities with no access to landline phones to become connected to the rest of the world, cloud computing can enable disadvantaged communities to leapfrog into the next generation of information technology. In order to fully exploit the benefits of this new IT model, the development sector will need to invest in training local stakeholders with the skills and expertise needed to take advantage of this new IT model, which will also have a positive impact on the broader knowledge base in developing countries. Taking cues from successes with consumers and businesses, it is possible to contemplate how cloud computing could be exploited to make a difference in the development context.

Healthcare. Advancing the quality of healthcare is a key development objective. In fact, three of the eight MDGs adopted by the United Nations pertain to healthcare combating HIV/AIDS, malaria and other diseases; improving maternal health; and reducing child mortality. In developing countries and rural areas, a high percentage of healthcare complications and fatalities arise out of medical errors, misdiagnoses and the lack of basic knowledge and expertise. One objective of healthcare NGOs in developing countries is to improve the level of expertise among medical professionals serving these communities. However, despite the best of intentions from all sides, a practical challenge that these organisations face is the difficulty of brokering exchanges and communications among interested parties—those looking for assistance and those willing to provide it. While it is common practice for doctors in developing countries and rural areas to refer to health information websites, these doctors also require a knowledge-sharing service in order to tap into the expertise of their peers and top medical professionals from around the world. This would take the current practice of telemedicine to the next level, creating a network that goes beyond the one-to-one, patient-to-doctor or doctor-to-doctor interactions. The Internet, as an open, global communications network, provides a mechanism to facilitate such exchanges. But that is not sufficient without the necessary IT solutions and systems—such as a database of experts categorised by medical specialty, a communication forum to post inquiries and address questions to specific experts and a searchable repository of previous inquiries. These are not trivial tools or solutions to develop. However, with cloud computing services, a development agency can put together all the pieces necessary to get a system of this nature up and running in relatively short order and with a relatively modest upfront investment. In this way, a medical professional in a village in Bangladesh, who may have a patient suffering from an infectious wound, could instantly correspond—possibly on a mobile device—with other doctors within the region and outside who may have more experience with such a case. Another challenge in the healthcare sphere in rural areas is ongoing patient care. With few clinics and limited medical staff, healthcare providers often lack the means to supervise treatments and monitor patient progress. This was, for example, a major roadblock when the South African government formulated a policy in 2004 to administer antiretroviral drugs to all HIV patients who had developed AIDS. Implementing this policy required health authorities to be able to track drug regimens and monitor the effect of the drugs on each patient. In order to do this, the International Development Research Centre (IDRC), in partnership with local South African organisations, funded the deployment of an information technology system. Using this system, staff in clinics can now enter patient data on computers or hand-held devices. These electronic medical records are sent daily to a central location where patients are monitored for resistance to the drugs. The system, which also gives clinicians reminders for patient care, has played a key role in the South African government's AIDS programme. Similar IT systems serving the purpose of managing ongoing patient care would be tremendously valuable in rural clinics around the developing world. But most such clinics do not have the resources or funding to build, maintain and manage IT systems. With cloud computing, it is possible to have patient care systems—similar to the one funded by IDRC in South Africa—implemented widely, but without requiring clinics to buy and manage all the hardware and software. Incorporating such solutions into the operations of clinics is one way in which healthcare NGOs can advance the quality of patient care in rural communities.

Microfinance. Development organisations have recognised that an effective way to address extreme poverty—another one of the MDGs adopted by the United Nations—is through injecting a business mindset into local communities. Thus, over the last decade, microfinance has proven to be a great catalyst for small business entrepreneurs in developing countries. Despite the buzz around microfinance, the reality is that there are significant challenges in scaling this financing model. Given the very nature of lending 'micro' amounts, the cost per loan is often too high for this model to scale broadly. The process of screening potential clients and processing loans is a cumbersome task. Moreover, it is difficult for microfinance institutions to follow consistent standards in granting loans, which has a direct impact on the ultimate success of their programmes. And once loans are made, microfinance institutions have a hard time managing their portfolio of loans—tracking collections, monitoring overdue accounts and making sure loans are used appropriately. Though microfinance works largely on personal connections and relationships in local communities, loan officers still need a way to administer their operations and report back to their sponsors, such as NGOs, credit unions or financial institutions.

Currently, the technology employed by many microfinance institutions, especially smaller ones, is limited primarily to using spreadsheet programmes. As a general rule, these institutions do not have the resources to deploy sophisticated IT systems similar to those employed by commercial lenders, yet they have to manage complex operations. Again, cloud computing can enable microfinance institutions to utilise easy-to-deploy IT solutions that create efficiencies and lend transparency to their financial management and performance. Imagine, for example, if a loan officer at a small microfinance agency in Africa could simply go to the Internet and log in to a website to screen loan applications, manage existing loans and track collections. And imagine if the executive director or sponsor of the microfinance agency could go to the same website and log in to his or her own account to view the agency's aggregate loan performance as well as the performance of each region, office or loan officer. A few microfinance institutions in Ecuador, Nicaragua, Honduras and India have begun to experiment with some of these IT concepts into their day-to-day operations. Though microfinancing is largely a private undertaking not controlled by any one organisation, it would be in the interest of microfinancing associations and development agencies to build and promote IT solutions—based on cloud computing—that could be adopted more broadly in the microfinance world. Using these solutions, microfinance agencies will be better equipped to scale and help alleviate extreme poverty in more parts of the world.

Disaster recovery. While disaster recovery is not directly a development objective or one of the United Nation's MDGs, the increased incidence of hurricanes, earthquakes, wildfires, floods and other disasters is displacing communities and causing tremendous damage throughout the world. Hurricane Katrina in the United States was a reminder that no community—including those in wealthy nations with substantial resources at their disposal to manage emergencies—is immune to the logistical nightmare of major evacuation and recovery operations. Information technology can serve as a critical tool in evacuation and recovery operations, but there is little luxury of time to develop and deploy systems. Given its turnkey nature, cloud computing can make a significant difference in mobilising resources in emergencies. In fact, when the official response to Hurricane Katrina was deemed sorely insufficient, a number of grass-roots and community-organised initiatives sprang up. These efforts were able to leverage the power of Web-based services to get up and run quickly and with minimal resources. One such initiative was the creation of an online database—running entirely on shared infrastructure—for evacuee and survivor tracking. This effort enabled families and friends, dispersed across many cities and states, to identify and locate one another or to provide clues that could help in locating missing victims. Another initiative, named the Broadmoor Project, set up a database of over 2400 homes located in the Broadmoor section of New Orleans. The purpose of this initiative was to match available resources and volunteers to houses that needed repairs. Using this service, the Broadmoor community was able to effectively tap into the thousands of volunteers who came to New Orleans with a commitment to rebuild the city. These initiatives were extremely effective. However, what were scattered, grass-roots efforts during Hurricane Katrina should become a regular and routine part of evacuation and disaster recovery operations—not just in the United States, but when natural disasters hit poor countries. A number of NGOs involved in disaster recovery, including the Red Cross and the United Nations World Food Programme, are already piloting IT solutions running on cloud computing infrastructure to help with mission critical activities such as procurement and distribution of food and supplies during natural disasters. As emergency management agencies and NGOs work together to build a best practices blueprint for managing disasters, they should standardise on such solutions that can be deployed instantaneously when disasters occur.

9.17 Summary

In Indian market cloud computing holds immense promise. At present, in nascent stage, the Indian market has a huge, untapped potential at every level, public, corporate and IT. It is the ideal test bed to pilot strategic adoption technique as it gives a platform vendor access to the whole ecosystem. It provides one with gamut of opportunities to try and experiment. Even Microsoft has acknowledged the future of cloud computing in India by saying that India has a huge potential to emerge as the global competency centre for cloud services. Since the inception of cloud computing services, the IT industry has seen a dramatic change in the business processing. Even organisations like Microsoft are now adopting the cloud computing technology to get a bigger market share for business processing outsourcing. Cloud computing service providers are over enthusiastic and are ready to grasp the share of cloud computing business. The governmental circles are equally positive and enthusiastic about cloud computing.

Considering the booming of small and medium enterprise which has largely been emphasised in CK Prahlad's 'The Fortune At The Bottom Of The Pyramid', cloud computing is the perfect solution for the existing time. Some of the inherent problems that India has been grappling with can now turn into a great opportunity for cloud vendors. Clusters of small businesses can be empowered through the cloud and in turn open up avenues of opportunity for cloud vendors. Not only the SMEs but any other form of organisation that cannot afford to invest a fortune to get their desired services can now hit their target in the form of cloud services. For example, an academic institute can subscribe to cloud services that provide student/teacher/parent collaboration on subscription. Thus one can safely say that cloud services are the need of Indian market today.

However one must allay his enthusiasm and consider the problems and obstacles in Indian market too. It is then when we realise that not everything is as easy and comfortable as is portrayed by governmental officers or cloud computing service providers. Techno legal experts do not seem to give their green signal to cloud services in India. Raising their point against such use, they say that the Indian market is not yet totally ready for such services as there are inherent weaknesses of Indian cyber laws, privacy laws and data protection laws and defective e-governance and ICT policies.

An Asian Cloud Computing Association, a tie-up association of vendors, has been formed to meet the challenges of cloud computing in Asia. This has been formed after realisation from vendors that the new opportunities would also bring huge problems and obstacles with it.

Thus one can say that India is still not ready for cloud computing. In the absence of regulatory and legal regime, it would not be wise to open the market to cloud services. Thus, proper legislations must come in place as early as possible to explore the huge potential offered by cloud computing.

Chapter 10
KM in the Development Sector

10.1 Background

Knowledge management is the concept under which information is turned into actionable knowledge and made available effortlessly in a usable form to the people who can apply it. Knowledge is insights, understanding and practical know-how that a member in an organisation possesses. Knowledge management is a collaborative and integrated approach to the creation, capture, organisation, access and use of an enterprise's intellectual assets.

Types of knowledge:

1. *Tacit*—it's a basic form of knowledge which is person centric where the person's cognition is knowledge. The cognitive skills as how he came out of a situation, how to interact with a set of people and how to get your work done in an environment constitute tacit knowledge. It's very subjective, hard to measure and situation and environment specific. The challenge of KM is that with employee moving out this information also moves out of the organisation and could not be used any further, and for completing the same task, the other employee or any other person has to go through the process of learning and unlearning which takes a lot of time. As time is money and in the coming age knowledge is going to be the only edge, it becomes very important for organisations to create such a robust system that it captures all the knowledge keeping it clutter-free with appropriate filters.
2. *Explicit*—the tacit knowledge is available to a person and could not be used by anybody else till the person who owns the knowledge shares it. Besides, using the knowledge by a larger set of people is also very difficult, as there is a limit to which a person can interact with other persons personally. So, to make the tacit knowledge concrete, we have to put it in the form of articles, data and other hard forms, where it can be stored physically and could be retrieved. The knowledge should be arranged such that it should be available to all the members of the organisation, all the time in an easily accessible way by not wasting time in finding the required information, only not the heap of information.

© Springer International Publishing Switzerland 2016
S. Mohapatra et al., *Designing Knowledge Management-Enabled Business Strategies*, Management for Professionals, DOI 10.1007/978-3-319-33894-1_10

The KM system should also incite trust in the system, and it should be non-partisan and integral in values which it eschews.

At times people misunderstood knowledge management with the tools that knowledge management uses. These tools are personal like notes, folders, diaries, to-do list, etc. and organisational like metadata. Knowledge management is a multidisciplinary branch which is a confluence of cognition, behaviour, analysing information, storytelling, languages, organisational behaviour, note-taking aptitude, etc.

10.2   Need of KM in Developmental Organisations

The developmental organisations are no more different than corporate; rather, it's very crucial for developmental organisations to capture the knowledge effectively. Their employee base is sometimes bigger than corporate; like in India, Self Employed Women's Association (SEWA), an organisation of poor, self-employed women workers who earn a living through their own labour or small businesses, has got an employee base exceeding 100,000 and BASIX >25,000. So, with this huge employee base, they need to manage efficiently the knowledge captured by their employees. Moreover, it's far crucial for a developmental organisation to capture the knowledge because they are working directly with people. Any knowledge pertaining to people, their behaviour, culture, customs, demography of the community and the area the community is living is very important and takes a lot of pain to be collected. If the knowledge is not captured efficiently and fully, the organisation may face a problem. As the attrition rate is high and increasing, in this scenario, effective knowledge capture becomes more crucial. Any employee, who has worked on ground level, will take huge knowledge out with him/her. The biggest capital a developmental organisation has is the goodwill in the eyes of the people; when an employee with goodwill moves out of the organisation, it's a substantial loss of capital for the organisation not in terms of human resource but in terms of precious goodwill. As goodwill building takes a long time, it would not be possible for an organisation to fill the void for a long time and hence it suffers.

At the same time, the KM systems should be simple, easy to understand and cost effective. As the development organisations are struggling for funds to continue their activities, KM should not put extra pressure on them in terms of resources and time of the employee. It should also be built in the system of the organisation to capture the knowledge. Knowledge management system should not work as a standalone unit; in that case, it would not be effective until it is linked to the key areas of performances (KFAs) of the employee. When it is linked with KFAs, it is ingrained in the system and hence needs not any extra efforts and will be completed in time.

10.3   Knowledge Products

Nowadays, no service is purely a service and no product is purely a product; there is something in common between them. The same happens with the products or services offered by organisations. Any service rendered by organisations contains part of knowledge, e.g. in an area where a developmental organisation is working for livelihood promotion, the choice of activities it undertakes (like agriculture, not animal husbandry) contains an element of knowledge; it is based on historical experience of the developmental organisation that this kind of activities is not sustainable or the market of these types of products is not existing. This could also be observed in patterns on how the community members are organised like somewhere in co-operatives or self-help groups (SHGs), depending on the historical experience of what goes well with the community members. All these information are unique to the ground-level workers who are working there for a long time, and the organisation has been able to capture the knowledge acquired by the employees.

10.4   KM Cycle

KM cycle is a process of how information changes from tacit to explicit in an organisation and makes a valuable asset in the form of reusability of the information. There are various KM cycles proposed by different workers. Typically a knowledge cycle starts with identification of the knowledge which is useful and should be captured, and then it moves through various stages of codification (where tacit knowledge is changed into explicit), validation and integration based on the different KM cycle models (designed by various knowledge management experts).

10.5   Concept of Trust in Development Organisations

Trust is most crucial in sharing information. When the employees don't have trust on other persons or in organisation systems and management, they may not be willing to share the knowledge, as knowledge is the most powerful tool after information and parting or sharing the knowledge may seem losing the edge to the employees. Trust also forms an important part in accessing the information by the employees. While retrieving the information from the KM system, the employee should have faith that the information is correct and he/she does not require to look or refer for any other source. Trust also constitutes an important part on the part of an organisation; it should not let the trust level of employees go down by any inadvertent exercise by any employee where he/she can use the information for any ulterior purpose. Trust development takes a long time and right environment. It has to be built over years and get it stronger. Making trust takes a long time but breaking trust takes a moment. Any breach in trust will make all efforts till the date of incidence to be null.

Trust in development organisations is the only capital on which the whole structure of the organisation stands. As any organisation starts working in an area, it grows by building trust. Trust development is an ongoing exercise; it's not a one-time exercise. Development organisations should not be allowed to breach trust as its mandate. Even if an employee is caught at the wrong foot and he has shared this information to the organisation, then he must have trust in the organisation that this information of mistake will not be used against him in the future. This kind of incidence cements the trust of the employee in the organisation. Organisation should not lose any chance of cementing trust between the management and employees and between employees.

Trust also enables the organisations to collaborate horizontally with other organisations. As in today's scenario, where the consortium approach is common, many donor agencies finance a project, or a single donor agency involves a large no. of partners in a single project. In these kinds of cases, it becomes very difficult for the organisations which are financing the project to ascertain the level of trust in different partners and work accordingly. The partners should have trust in competence, co-operation and accountability of the partner organisation that it is collaborating. It could be achieved by sharing the crucial information and aligning the KM systems of the project with the requirements of the partners. There could be formal agreements between the organisations which are collaborating or which are partners, but lack of trust cannot be bridged by formal agreements as agreements cannot cover all the terms and conditions arising in the future, though formal agreements could be the first step in concertizing trust which over a period of time creeps from papers to the practice and become norms of collaboration.

10.6   Individual-Level KM

There are many crucial individual-level skills that form part of KM, and it should be made clear that these skills are the means or tools which every employee should possess to contribute towards KM system to work effectively. These basic personal characteristics include:

- Ways in which people access information out of a heap
- Writing notes to capture the essence at the spot
- Extensive reading habit
- Knowledge creation by drawing conclusions and inferences based on previous actions and results
- Interpersonal communication

These basic characteristics are shared by every employee in the development sector but they are not conscious. As we have stated earlier, there should be incentives, and workers should be made aware and sensitise towards the need and importance of capturing knowledge. Developmental organisations should device innovative methods to inspire and motivate people to capture the information, analyse, make inferences and record it in a database where it could become a part of the organisational memory and could be retrieved as per need. At this stage, this factor must also be paid attention that not only knowledge be captured, but it should be arranged in such a manner that it should be retrieved with least efforts and should be relevant to the seeker.

While designing a KM strategy, it should be always kept in mind that the KM strategy is for the organisation and not organisation for KM strategy. It should be in sync with vision, mission and goals of the organisation and should include all the aspects of the organisational learning (like organisational culture, people's behaviour) in a holistic manner.

10.7   Transforming Personal Knowledge to Organisational Knowledge

Personal knowledge conversion to organisational knowledge requires a framework that enhances the capture of knowledge with acknowledging the originator. The system should be such to develop faith in the KM system. It should be participatory and ensures equal participation of all the stakeholders. The tools used for the KM systems should be such that they are easy to use for the employee, less time demanding and cost effective. It should not be such that one has to sit in front of the computer at the headquarter to enter the knowledge from a remote village; it could be made offline with hard copies also like writing a term paper every week or fortnight or month. Effective KM also incorporates CoPs (communities of practice) which are informal (most of the time) or formal (top-down approach) to act as a potential generator for the knowledge. In a developmental organisation, these are very important. There are some core activities of KM which are common to both corporate and any other organisation. These core activities can differ demanding on the mandate of the KM strategy of different organisations, but the basic purpose remains the same. These are as follows:

1. Knowledge Mapping
     Knowledge mapping basically means mapping of the knowledge, i.e. the origin, flow and end use. It means in detail the organisation has to describe where the knowledge starts, who and what captures and how one captures it and then flow that means how it is changed from tacit to explicit, who and how will it be done and then how it is stored and where, how and who can access it. Knowledge mapping can show how the knowledge is being collected right now, if there are any changes required to incorporate to make it more wholesome, who are the persons who are potential hurdles to the flow, what changes an organisation should go through, what are the current processes which require the changes and how much. It could also help to assess whether the organisation is leveraging its knowledge or not and if yes to what extent.
     When an organisation starts leveraging, its knowledge is to be made sure that it's a multidisciplinary approach with multiple stakeholders and all should be consulted and should be made a part of the larger frame. In this framework,

everybody's concern is taken into consideration like on employees' side; it should be less time-consuming, should not affect the core work and should be linked with performance appraisal, while on the part of the organisation, the need, strategy and cost-effective tools.

2. Goal Setting
   It occurs at three levels:

   (a) *Individual level*: While working in a knowledge-intensive environment, the employees should identify their knowledge management perspective and should align it with the organisational perspective and should decide sync, to go hand in hand with the organisation. In such multi-stakeholder exercises, they should freely air their concerns and should come to know that many other people share the same concerns; this will develop the trust to share knowledge and will help in alignment of the employees' personal goals with organisational vision, mission and goals.

   (b) *Organisational level*: At this level when employees have aligned their goals with the other employees, now the employees with a similar set of goals align their goals with the organisational goals. The management of the organisation should actively take part in this process and make sure that all the concerns of the employees should be taken care of.

   (c) *Partner level*: When an organisation has aligned their goals with employees, now they need to expand it further to the partners. It's like a phenomenon known as collaborative planning in operations area of supply chain management where all the partners sit together and decide their common goals. The goal is large enough to encompass each and every stakeholder's personal goals. In this way, the development organisation develops a common goal which is achieved in collaboration with partners.

3. Tools and Processes
   Selection: There are many tools available which fall basically in the domain of the communication tools. But these tools need not be expensive and complex. These tools can be as simple as like journals, magazines, notes and case studies to as complex as MS SharePoint, email newsletter, videos, conferences, workshops, etc. The need and fit of the tools should be finalised by the employees only; it should not be thrust upon them by the organisation in which case it might be rejected. Employees should also be demonstrated with the use of tools in a live manner.

   Implementation: When the tools are decided by the employees themselves, there are very little chances of rejection, and the tools evolve over a period of time as employees consciously take decision to modify them according to their need.

4. Monitoring and Evaluation
   This forms the next most important constituent of the KM strategy. There should be proper monitoring to avoid duplication and plagiarism. It should also be finalised by the employee only and should be finalised by peer review. In this way, it won't have any pressure on the organisation and makes this process more transparent.

10.8   Outcomes of a Successful KM Plan

There should be some matrix to evaluate the benefits accrued to the organisation with implementation of the KM plan. The matrix should not also take into account the tangible and quantitative benefits but also intangible and qualitative benefits. These outcomes should not only take into account the benefits accrued to the organisation but also the benefits to the employees. The benefits could be enlisted as:

1. *Making knowledge power visible*: From processes improvement to change in behaviour and performance of employees and the organisation as a whole, the KM exhibits its power.
2. *Common platform for discussion and planning*: KM should bring all the stakeholders to the same platform to discuss and plan their future actions in sync with each other.
3. *Participatory decision-making*: A common platform enhances the ability to make decisions involving larger stakeholders. After KM implementation, no employee can say that he/she does not have the knowledge of the process; it makes everybody in an organisation empowered and makes the organisation democratic.

10.9   BASIX: A Case Study of Knowledge Management

10.9.1   Background and Working Environment

BASIX, a livelihood promotion organisation, was set up in 1996 by Mr. Vijay Mahajan; he is currently serving as a Chairman. It is working with more than 1.5 million customers serving 90 % of rural households and 10 % urban slum dwellers. The organisation has stretched its arms to 16 states of the country.

A development organisation's basic aim is to provide various sustainable livelihood options to the rural poor and women by creating a holistic approach to its working and to have an integrated approach that will include financial assistance as well as technical assistance so that they can employ the funds in the most productive manner. To fulfil this goal, the services provided should include not only financial services but also agriculture, livestock and enterprise development services and institutional development services. It isn't just about giving loans but also about creating livelihood mechanisms, which would build capacity among the poor to repay their loans easily and leave them better off than before.

BASIX is such organisation which is building its human capital so that there is a strong trust built in the minds of customers and they should voluntarily ask for assistance. Employee turnover is reported to be even less than 1 %.

Every employee who is supposed to work with the organisation for a sufficient period of time has to undergo a 3-month training programme. During the induction period of 3–4 weeks, the candidate has to imbibe the learning about the organisation and working of its various departments. For a development organisation carrying out loan as well as technical assistance, activity field visit is bound to be a daily course of its life. During the induction and thereafter, the person is put into a rigorous orientation programme to give him/her a role clarity in the organisation.

The employees are brain ware for any organisation. They are the biggest source in terms of providing managerial training to their subordinates and advice to senior level office. The organisation has no selfish motive of retaining an employee. The organisation has no expectation that its staff would be working in the same organisation for his or her complete working life.

10.9.2   Knowledge Management in BASIX

BASIX is working on a livelihood triad based on three core activities:

1. Institutional Development Services (IDS)
2. Financial Inclusion Services (FIS)
3. Agri Business Services (AgBDS)

10.9.3   KM Process Followed in BASIX

The instrumental knowledge is put in practice by various educational programmes, and the practices are reflected by livelihood learning groups which in turn by research and publications are out in theory which through process of manual development is again converted into instrumental knowledge, thus competing the cycle of KM process (Fig. 10.1).

10.10   Conclusion

The development sector has a huge latent demand for KM. Though there is an eminent need, they are not aware of this kind of solution, so they need to be made aware and sensitise about the issue. There is a certain development organisation who understands the criticality of such an initiative, but they fear the heavy cost of such initiative on lines of their corporate peers. Here, these types of organisations should be told that they could design organisation-specific KM system depending on the budget and specific requirements the organisation has. An intelligently and participatory designed KM system could answer many problems and prove to be a cost-cutting measure.

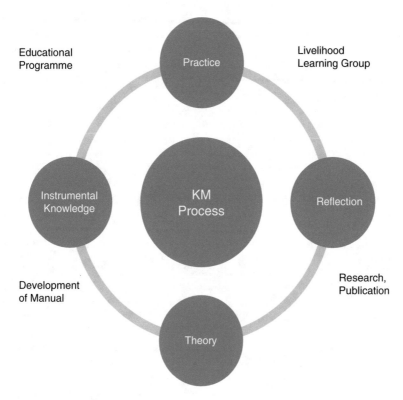

Educational
Programme

Livelihood
Learning Group

Development
of Manual

Research,
Publication

Fig. 10.1 KM process followed in BASIX

10.11 Summary

The chapter discussed the need for knowledge management in the development
sector. It also discussed different products available for development sector which
can help increase reuse of knowledge. To capture knowledge, KM cycles are
followed. The cycle helps to capture knowledge at the individual level, then at
group level and finally at organisation level. Finally, knowledge cycle at BASIX has
been explained.

Chapter 11
Automation in Knowledge Management

11.1 Introduction

The impact of globalisation, free-market economy, current economic conditions, organisational downsizing, crunch of skilled and talented employees, etc. has increased the recognition of the importance of managing the corporate knowledge. Knowledge gives us, as individuals, the power to grow, to better ourselves and to succeed in our endeavours. In the business world, knowledge is power as well. As with individuals, knowledge is what enables businesses to grow and to succeed. Contained within every business is a wealth of knowledge about its policies and practices and about the industry that it serves. A business uses this knowledge to sell and support the products and services that it offers. However, knowledge is not held within a business itself, but rather by the various individuals that make up the organisation. Without the knowledge possessed by the people within it, a business cannot survive. To a business, therefore, knowledge can be considered an asset—perhaps the most important asset that it has.

11.2 Technology and Knowledge Management

The power of technology in supporting knowledge management (KM) activities is widely recognised. However, in most KM literatures, the discussion on related technology is either given cursory treatment or confined largely to product-specific features. This reflects a division within the KM community. One camp is represented by consultants whose paradigm is rooted in concepts such as organisational learning and organisational memory. They tend to view KM as a strategy and often treat technology as a 'black box' because of the intricate technicalities involved. The other camp is represented by technologists, who tend to be product centric and focus on features and functionalities of the systems. They perceive technology as the primary solution to resolve KM issues.

© Springer International Publishing Switzerland 2016
S. Mohapatra et al., *Designing Knowledge Management-Enabled Business Strategies*, Management for Professionals, DOI 10.1007/978-3-319-33894-1_11

As a result, a KM practitioner who is not technically informed but wishes to use technology as part of a KM implementation would have difficulty in selecting from a vast array of technology solutions. Conversely, a KM practitioner who is very familiar with specific technologies but ignorant of KM processes could recommend solutions that may not meet the needs of the client.

11.3   Definition of Knowledge Management

Knowledge management is a newly emerging, interdisciplinary business model dealing with all aspects of knowledge within the context of the firm, including knowledge creation, codification, sharing and how these activities promote learning and innovation.

We extend the above definition to the knowledge management systems (KMS) as follows:

Automation in KMS is all about having systems and technologies in place to automate the enterprise-wide processes, methods and techniques of learning organisations enabling them to manage their knowledge assets.

11.4   Factors That Affect Knowledge Management

11.4.1   Economics

If organisations can manage the learning process better, then they can become more efficient by converting tacit understandings to effective knowledge management. Developing these learning strategies has subsequently become an important knowledge management theme.

Another driver of knowledge management that comes from economics is how to account for the significant performance variation. Customers expect the same level of consistency, reliability, responsiveness and quality across the globe for any organisation.

11.4.2   Sociology

Sociology has contributed both macro and micro perspectives to knowledge management. At micro level, sociology has strong research interest in the complex structures of internal networks, and communities have obvious relevance to knowledge management. Knowledge management has inherited that concern for social facts. Rather than build from theory, it looks at what people actually do, the circumstances in which they share knowledge or do not share it and the ways they use, change or ignore what they learn from others. Those social facts guide the development of knowledge management tools and techniques.

11.4.3 Philosophy and Psychology

Almost from the beginning, knowledge management has explored the differences between tacit knowledge and explicit knowledge and between know-how and know what. In recent decades, burgeoning electronic information storage has made access to vast quantities of information given in developed nations. This has highlighted the importance of the undigitised knowledge. The value has two sources: one is scarcity—the value of the expertise that cannot be imitated and is widely accessible—and the other is the role of that knowledge in organising and selecting from the flood of information so that it can be put to use.

Taken together, the conceptual rigour of economics, the observational richness of sociology and the understandings of philosophy and psychology give knowledge management the intellectual scope and substance it needs to wrestle with the real human and structural complexities of knowledge in organisations.

11.4.4 Practices

The three practices that have brought a lot of relevance to knowledge management are information management, the quality movement and the human factors/human capital movement.

Information management that was developed during the 1970s and 1980s is usually understood as a subset of the larger information technology and information science world. Information management is a body of thought and cases that focus on how information itself is managed, independent of the technologies that house and manipulate it. In broad terms, knowledge management shares information management user's perspective—a focus on value as a function of user satisfaction rather than efficiency of the technology that houses and delivers the information.

The quality movement focused on internal customers, overt processes and shared, transparent goals. Quality techniques were applied most successfully to manufacturing processes, while knowledge management has a broader scope, including processes that do not seem to lend themselves readily to measurement or even clear definition.

By definition human capital focuses on individual, whereas most knowledge management work is concerned with groups, communities and networks. Nevertheless, knowledge management builds on human capital ideas and has, as one of its tasks, to continue making the value of human capital clear to organisational leaders while developing tools and techniques for investing and reaping benefits from it.

11.5   Laws of Knowledge Management

The basic self-explanatory laws of the knowledge management are:

1. Knowledge is the key to business survival.
2. Communities are the heart and soul of knowledge sharing.
3. Virtual communities need physical interaction.
4. Passion drives communities of practice.
5. Knowledge sharing has an inside-out and an outside-in dimension.
6. Storytelling ignites knowledge sharing.

11.6   Approaches to Knowledge Management

11.6.1   Mechanistic Approach

This is characterised by the application of technology and resources to improve the status quo. Following assumptions are required for the mechanistic approach:

1. Better accessibility to information along with enhanced methods of access and reuse of documents.
2. Networking technology and groupware would be key solutions.
3. Technology and sheer volume of information are crucial.

Although such an approach is easier to implement because of the familiarity of the technology and techniques, but unless the knowledge management approach incorporates methods of leveraging cumulative experience, net result may not be positive.

11.6.2   Cultural/Behavioural Approach

The roots of this approach are in process re-engineering and change management, and it views the knowledge problem as a management issue. This approach focuses more on innovation and creativity than on leveraging existing explicit resources or making knowledge explicit. The following are the assumptions for the cultural/ behavioural approach:

1. Organisational behaviours and culture need to be changed for achieving business objectives.
2. A holistic view for problem-solving is required.
3. It is the process that matters and not the technology.
4. Nothing changes unless a manager makes it happen.

Cultural factors affecting organisational change are often undervalued, and the cause-effect relationship between cultural strategy and business benefits is not very clear.

11.6.3 Systematic Approach

It has its roots in rational analysis of the knowledge problem and applying newer techniques of solving the problem. The following are the assumptions for systematic approach:

1. Sustainable results matter more than any process or technology.
2. Modelling is critical for resource handling and managing organisational knowledge.
3. Solutions can be found in a variety of disciplines and techniques, and traditional methods of analysis can be used to re-examine the nature of knowledge work.
4. Change of policies and work practices are essential in incorporating change management.
5. Although knowledge management should be championed by managers, it is just not the sole responsibility of the managers.

11.7 Business Mapping of Automation in Knowledge Management Systems

In the past, computers, specifically mainframes, mainly performed algorithmic bookkeeping operations, rather than decision support. Now that networked terminals and personal computers are a fixture on every front office desktop, computers are mainly being used directly by knowledge workers themselves. The proliferation of computer technology makes a company's information technology infrastructure the ideal place to adopt knowledge management principles. This means that businesses need to implement systems and applications whose behaviour is determined primarily by the company's knowledge assets, to assist, augment and support workers' decision-making efforts.

Unfortunately, most applications are not knowledge automation systems. Traditional applications are created using procedural coding techniques, wherein knowledge is hard-coded into the application as a series of step-by-step procedures. In this model, although knowledge is being managed and utilised, it is contained in an inflexible, static structure. The behaviour of the system is driven by the procedural code, while the decision-making efforts of knowledge workers are guided externally, resulting in a fundamental separation between the knowledge worker and the system. This relationship is illustrated in the following diagram (Fig. 11.1):

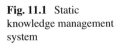

Fig. 11.1 Static knowledge management system

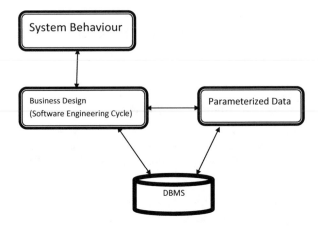

In this type of application, the separation between the knowledge worker and the system means that a company's business model is applied inconsistently. Also, because the behaviour of the system is determined by procedural source code, changing that behaviour requires changing the code, which is an error-prone, time-consuming and expensive process. Finally, since industry change occurs more rapidly than applications can be updated, the software engineering cycle cannot keep up with the evolution of the marketplace.

In a true knowledge automation system, knowledge assets drive an application's behaviour, rather than procedural code. Knowledge automation systems incorporate expert system technology, which replaces procedural code in an application with a knowledge base, to contain knowledge, and an artificial intelligence (AI) engine, to drive decisions. This model allows the knowledge base to be updated and maintained by knowledge workers, closing the gap between the knowledge worker and the system, as illustrated in the following diagram (Fig. 11.2).

With knowledge automation systems, the business design defines the behaviour of the system, allowing changing usefulness needs to be implemented directly through the knowledge base, rather than through the software engineering cycle. This means that the system's behaviour can change dynamically with the business, resulting in a dramatic reduction in turnaround times for meeting new business objectives. This, in turn, results in a more responsive organisation that is able to maximise the use of its knowledge assets for tangible benefits. Implementing a knowledge automation system requires a change in mindset—IT departments must learn and embrace rule-based programming in order to provide a knowledge automation system's infrastructure. This requires the adoption of expert system technology, to enable the creation of a knowledge base and the use of an AI engine. Just as people apply intelligence to knowledge to make decisions, so does the knowledge automation system apply artificial intelligence to the knowledge base to determine the system's behaviour. Once these components are in place, the system can be easily maintained and updated through the combined efforts of both knowledge workers and IT personnel.

Fig. 11.2 Knowledge
automation system

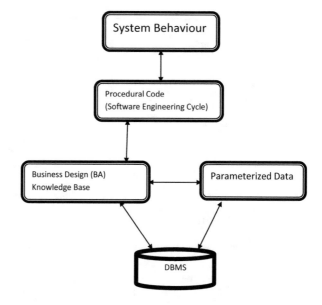

11.8   Knowledge Management Process Model

The process model can be separated into two major parts: the coordination processes and the operational processes. The coordination processes represent the management tasks related to KM; these include analysing and planning KM, dealing
with organisational issues, etc. The operational processes present the processes of
actually carrying out KM, i.e. knowledge collection, sharing, update, etc.

The coordination of KM activities is the very centre of all other activities as
everything is initiated and controlled from here. Knowledge is always already dealt
with (created and also shared), although the necessity to manage these processes
might not have been realised yet. This clearly indicates that it is a question of
improving current practices instead of replacing them with some entirely new ones.

First, the existing knowledge stacks, the channels used to transfer knowledge and
the general surroundings have to be analysed to enable their management. Then a
target state needs to be defined. To enable both an effective tracking of the success
and an identification of any shortcomings of the processes set up, it is necessary to
define metrics. This at the same time provides a possibility to make the usefulness
of KM visible.

The KM coordination processes are shown in the figure below (Fig. 11.3).

The general concept of the process model is that within the coordinating processes, the operational processes are planned and initiated. Together, these make up
the KM system. The main processes are described in the following. 'Identification of
Need for Knowledge' identifies a need for knowledge and determines it. 'Sharing' is
initiated in order to find out whether knowledge that already exists in the system can
be used. This covers both the searching for knowledge by a person who needs the
knowledge ('Knowledge Pull') and the feeding of knowledge to recipients who are

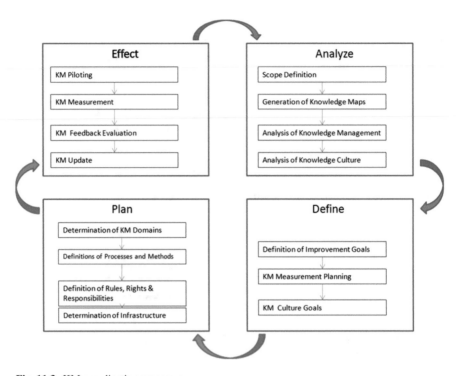

Fig. 11.3 KM coordination processes

known to be in need of it ('Knowledge Push'). If the needed knowledge is not available yet, 'Creation of Knowledge' is initiated. Consequently, the new knowledge (the result) has to be collected—this is done in 'Knowledge Collection and Storage'. Also when sharing knowledge, new knowledge is often created throughout the combination of the shared knowledge with the receiver's existing knowledge—this is indicated in the graphic by the activation of 'Knowledge Collection and Storage' by 'Sharing'. Both 'Creation of Knowledge' and 'Sharing' may have external links.

The KM operational processes are shown in the figure below (Fig. 11.4).

11.9   Knowledge Management System Architecture

The power of technology in supporting knowledge management (KM) activities is widely recognised. Here we describe a model of knowledge management systems (KMS) architecture for KM practitioners. The primary objective is to provide a framework for the review of technologies being used in supporting the fundamental KM processes. This attempt represents a modest step to bridge the gap between consultants and technologists in the KM community.

The three-tiered KMS architecture identifies three distinct services supported by KM technologies. They are infrastructure services, knowledge services and presentation services.

The three-tiered KMS architecture is shown in the figure below (Fig. 11.5).

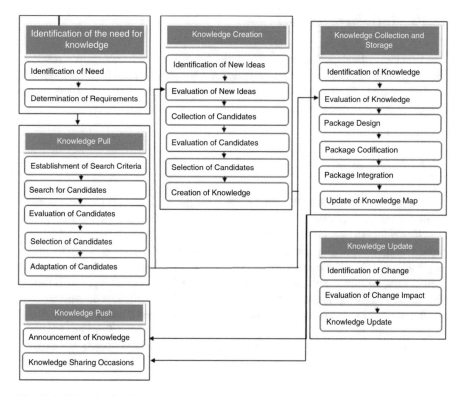

Fig. 11.4 KM operational processes

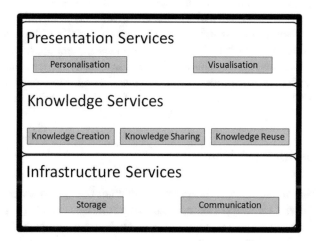

Fig. 11.5 KM system architecture

11.10   Infrastructure Services

The first tier in the KMS architecture model comprises the infrastructure services. Infrastructure services refer to the basic technology platform and features needed to implement KM. The two main infrastructure services provided by technology are storage and communication.

11.10.1   Storage

A technology-enabled store, known as a knowledge repository, is typically defined by its content and structure. The content refers to the actual knowledge stored. The structure refers to how each 'knowledge unit' is specified, the format in which it is represented, the indexing scheme and how each 'knowledge unit' is linked to others. In particular, a knowledge repository could either be populated with data or documents. Increasingly, repositories have been designed to capture graphical information such as engineering drawings and audio, video and multimedia documents. The trend is to develop repositories capable of supporting content that is less structured and of greater richness. Technology-enabled repositories also form the basis for supporting KM processes, particularly knowledge creation and knowledge reuse.

11.10.2   Communication

Increasingly, the shift is towards developing communication technologies capable of creating social presence and which possess multimedia capabilities. Social presence is the degree of salience of the other person in the interaction and the consequent salience of the interpersonal relationships. Multimedia capabilities allow rich content such as voice, images and video to be transmitted. Thus, the combination of social presence and multimedia capabilities provides a closer approximation to the actual face-to-face interaction. This trend gives rise to technologies that are designed specifically to support the process of knowledge sharing. More details about such technologies will be discussed in the next tier of the KMS architecture.

11.11   Knowledge Services

The second tier in the KMS architecture model comprises the knowledge services. Knowledge services are supported by technology solutions intended to help achieve the goals of KM directly. The three primary goals are to promote the process of generating new knowledge, encourage the flow of knowledge among organisation members and ensure the ease of access to knowledge repositories. The underlying knowledge processes of these three KM goals are knowledge creation, knowledge sharing and knowledge reuse.

11.11.1   Knowledge Creation

Knowledge is created either through exploitation, exploration or codification. Exploitation refers to the refinement of existing knowledge into new knowledge to achieve improvement in efficiency and effectiveness. Exploration refers to the creation of knowledge through discovery and experimentation. Codification refers to the articulation of tacit knowledge into formats such as formulae, manuals or documentation that are comprehensible and accessible to others.

Codification has been traditionally supported by technology through text documentation. Emerging technologies have augmented text documentation with other media such as digital video and audio. While non-text digital media are currently more difficult to search and browse than text documents, improvements have been made to facilitate browsing through video documents. A technique, known as summarisation, is used to automatically produce a gallery of extracted still, searchable images. Meanwhile, work is underway to improve the accuracy of automatic speech recognition (ASR) that seeks to support speaker-independent recognition with unconstrained vocabulary. The aim is to produce text transcription from digital audio to enable browsing and searching.

11.11.2   Knowledge Sharing

Knowledge transfer refers to the flow of knowledge from one part of the organisation to other parts. If this process is not properly managed, valuable sources of knowledge in the organisation will remain local or fragmentary and internal expertise underleveraged. Hence, one important goal of KM is to foster the flow of knowledge among organisation members. Technologies developed for the former purpose are known as social network analysis tools, while those developed for the latter are known as collaboration tools.

An emerging area in collaboration tools is social computing. Social computing is the development of digital systems that are drawn from social information and context to enhance the activity and performance of people and organisations. Its aim is to create social presence among the users. For example, Babble is a social computing system developed by IBM. Babble is similar to a text-based chat system but differs significantly in its ability to make the presence and activities of users visible through an interface known as a social proxy. Initial research at IBM showed that the daily interactions on Babble were able to engender relationships among its users.

11.11.3   Knowledge Reuse

The term 'knowledge reuse' in the KM literature is largely synonymous with 'information retrieval' in the information management literature. The process of knowledge reuse can be described through four main stages, namely, capturing knowledge, packaging knowledge, distributing knowledge and using knowledge. Two salient technologies are developed for knowledge reuse—content management and concept mapping.

Emerging technologies aim to provide enhanced search capabilities such as increasing knowledge of the user's needs and automatic generation of meta-data. Currently, most search systems gather information about users' needs solely from the query submitted. However, one study has reported that the average number of words in queries submitted to most Web-based search services is only 2.3. This information is apparently insufficient to build the context of the search or the profiles of the users. Personalisation services, discussed in the next section, seek to address this issue. The second area of research is related to meta-data generation. The value of meta-data is in encapsulating information about the document that can be used to construct selected views of the information based on the users' requirements. An example of such views is the thematic listing of documents. The future trend is for meta-data to be automatically generated through document analysis and classification.

11.12   Presentation Services

The task of ploughing through the increasing amount of information every day and integrating it from a variety of sources to support decision-making has proved to be daunting. Information overload is a syndrome that occurs when the quality of decisions is compromised as a result of spending too much effort and time to review more information than necessary. This syndrome is addressed to some extent in the final tier of the KMS architecture presented here, that is, by presentation services.

Technologies that provide presentation services are primarily concerned with enhancing the interface between the user and the information/knowledge sources. Two common features of presentation services are personalisation and visualisation.

11.12.1   Personalisation

Personalisation involves gathering user information and delivering the appropriate content and services to meet the specific needs of a user. It is implemented by aligning three components, namely, users' profile, content and business context. Users have profiles that represent their interests and preferences. A user profile is defined by a set of attributes and assigned values. Likewise, content is profiled on the basis of a set of attributes and values. The business context refers to the rules that determine how users and content are matched, based on their attributes and values. Personalisation solutions are rarely stand-alone but embedded in other applications such as the email. For instance, an intelligent agent is a type of personalisation solution that allows users to develop rules for automatically handling email messages, based on subject matter, source or other characteristics.

11.12.2   Visualisation

A second feature of presentation services is visualisation. The purpose of visualisation is to help users better understand the information and knowledge available by making subject-based browsing and navigation easier. Research which compared information retrieval from text with two-dimensional and three-dimensional interfaces found that richer interfaces provided no advantage in the search tasks that were studied. This finding could explain why the use of graphical visualisation is not as widespread as that of text-based interfaces in search applications.

11.13   Aggregation of Services

Even though the KMS architecture model distinctly illustrates the various services supported by technology, delineation among the services may sometimes be fuzzy. Several products have been developed to provide aggregated services. For example, knowledge manager from Microsoft offers multiple functionalities such as content management and deployment, search and delivery of filtered information; Primus Answer Engine and Primus eServer from Primus Knowledge Solutions allow users both to capture and share knowledge; Lotus Domino, from IBM and Lotus, supports collaboration, knowledge discovery and knowledge mapping.

The aggregation of services provided reflects a consciousness among software vendors of the need to build a unified KM platform upon which their suites of products can be designed to interoperate. By offering a comprehensive suite of solutions, software vendors seek to offer a one-stop solution to their clients and eliminate compatibility problems among various applications.

However, consultants must be warned against recommending a technology solution merely for its extendibility, comprehensive functionalities and technical features. A technology solution should be adopted on the bases of its ability to support the particular KM goal of the company and the extent to which the foreseeable usage pattern blends well with the organisational ethos. For example, if an organisation aims to gain knowledge from customers or create knowledge for customers, a technology solution that primarily supports the knowledge creation process is preferred to one that supports only the knowledge sharing process.

11.14   Calculating 'Return on Investment' on KM Initiatives

The battle for funding different projects within an organisation is extremely competitive, more so in these uncertain economic times. Any spending in the ITES must have a demonstrable impact in the corporate bottom line or top-line. According to Ernst & Young's Survey of Fortune 1000 IT Buyers 2002, 'a technology investment isn't going to make it through the approval process if it doesn't carry with it a reliable financial analysis showing the impact the investment will have on an organization's bottom line. The results from this survey make that abundantly clear' (Source: Ernst & Young).

11.14.1   The US Navy Metrics Model

The US Navy metrics model provides for a classification of both KM metrics and the processes to which they apply. The US Navy Knowledge Centric Organization (KCO) model groups metrics into three categories relating to the classifications of the characteristics they are intended to measure. These measures are primarily process focused. They are:

- Outcome Metrics: measure high-level or strategic characteristics of the organisation such as overall enterprise productivity
- Output Metrics: measure tactical or process characteristics such as time to solve problems
- System Metrics: apply to measurable aspects of the process itself such as number of queries

This model goes further to describe the primary classes of business objectives for a KM initiative. These classes are:

- Programme and process management (applies to organisational strategies and objectives)
- Programme execution and operations (includes topics such as collaboration)
- Personnel and training (addresses objectives such as learning and quality of life)

Classes of business objectives	Metrics categories		
	Outcome metrics	Output metrics	System metrics
Programme and process management	X	X	X
Programme execution and operations	X	X	X
Personnel and training	X	X	X

11.14.2   Prerequisites for Any Model

Analytical Rigor: An ROI is a dollar calculation. Therefore, factors considered must have an actual or assigned dollar value. It is essential that the data that is assembled for the ROI be thorough. It must include all factors. To ensure a comprehensive analysis, care must be taken to address all of the key components:

- People
- Processes
- Products
- Tools
- Resources
- Customers

    While measuring costs:

- It is essential that the costs assembled apply to the process that will change as a result of implementing the evaluated alternative/initiative.
- The factors must be consistent. The comparison must be structured to compare 'apples to apples'. For example, if the status quo process costs include licensing fees, then the costs for the alternatives being evaluated must also include licensing fees if they are applicable.

Calculating an ROI requires a complete understanding of how the process is currently being accomplished:

- It is fundamental to compare the investment initiative with the status quo. Therefore, considerable information must be collected on the current process.
- Ideally, information should be collected in the same cost categories as are applied to the investment initiative to ensure that like values are compared.

11.14.3   Cost/Benefit Model

Creating a complete picture of the value of a KM initiative requires that both tangible and intangible factors be included in the analysis. This is essential because many of the benefits of KM are not readily assigned a dollar value. Since ROI is strictly a dollar calculation, intangibles fall outside its domain. Therefore, the cost/benefit analysis is preferred when dealing with KM initiatives.

- *Tangible benefits*: Tangible benefits are basically those that can easily be tabulated or a cost readily assigned to.
- *Intangible benefits*: Intangible benefits, also called soft benefits, are the gains attributable to the improvement project that are not reportable for formal accounting purposes.

Typically, benefits are not included in the financial calculations because they are nonmonetary or are difficult to measure.

11.14.4   Tangible Costs

Typically, tangible costs and benefits may be categorised as:

- Purchase
- Development (including the transition to the new process/system)
- Implementation (including change management)
- Maintenance

11.14.5   Cost Characteristics

Each of the four previous categories of costs can be further described by the following cost characteristics:

- *Sunk costs*: Sunk costs are typically assembled in an analysis such as this to help *put the status quo alternative in perspective*. These costs *should not be included in the actual ROI* calculation because the only factors relevant to the decision about the new initiative are those that can be controlled from the present time forward.
- *Recurring and nonrecurring costs*:
  - Nonrecurring costs include the baseline and acquisition costs, deployment, telecommunications, development, integration, testing, training, installation, parallel systems, data migration, phase-out and other pure IT costs.
  - Recurring costs include maintenance costs, operations, upgrades, facilities, staff, ongoing training, contracts and equipment replacement.

- *Cost avoidance*: Term used to describe those costs that are avoided by selecting a particular alternative

11.15   Measurements

11.15.1   Measuring Tangibles

By definition, measuring tangibles is considerably easier than measuring intangibles. Key factors here are accuracy, thoroughness and consistency. The cost measurements under focus are:

- Process costs
- Process change costs
- Other related processes (rewards/recognition, training practices, etc.)
- Level of service
- Benchmarking methods

11.15.2   Measuring Intangibles

Converting intangibles to tangibles: The following few examples illustrate how seemingly intangible factors can be converted into tangible values by analysis and measurement.

- Customer satisfaction can be measured in the customer satisfaction measurement (CSM) method. If it is measured as a 7 on a 10-point scale before the initiative and a 9 afterwards, a relative relationship can be established. By further analysing customer satisfaction levels and keeping data on sales or use of the product, it is possible to establish a relationship between customer satisfaction and sales or use of the product. Therefore, a dollar value of the difference in sales or use can be applied.
- Employee satisfaction can often be translated into retention which affects training costs. Again, measuring before and after allows calculation of the change in training costs as a cost avoidance.
- Innovation can be counted as the number of new ideas. The cost savings or increased revenue of these ideas can normally be determined.
- Acceleration of the time for a new staff member to become productive can be measured. If a person becomes productive after 2 months with the new KM initiative rather than 3 months under the status quo, 1 month's productivity is gained. One can assign a dollar figure to the 1-month productivity.

11.15.3   Actual Intangibles

Those intangibles that cannot be converted to tangibles should be addressed in the benefit component of the cost/benefit model. Benefits can be analysed in three ways:

- List the benefits of each in columns for comparison.
- Consider the pros and cons of the alternatives.
- Rate the intangibles.

11.16   The Fusion of Process and Knowledge Management

While KM and process engineering were being evolved in parallel, there was no serious effort to fuse them into a consistent, holistic architecture. KM programmes over the past decade have focused on organising employees into communities of practice and building repositories of 'best' or proven practices. There was (and still is) a general lack of understanding of how valuable the fusion of processes and knowledge can be. The thought of actually taking the distilled knowledge and making it easily available to people executing the process was somehow overlooked. Employees would only stop to access the available knowledge base when the process execution came to a screeching halt due to an inability on the part of the employee to continue. Many times this would involve looking up information in an offline source like a procedure handbook or calling a friend who might know the answer. A major thrust of KM efforts in the past 5 years has been building these employee locators who could answer questions involving specific knowledge domains.

On the process side, until quite recently, application systems have been almost purely transactional by design. Users simply enter and retrieve data from a data store. There is no way of easily accessing offline knowledge. While most modern applications feature built-in help features and drop-down table lists, the kind of access to knowledge bases that would represent a fusion of process and knowledge management has historically been ignored. Even access to electronic document management systems has been mostly an offline event and has not been specifically linked to processes.

In recent times, there has been recognition of the need to merge process and knowledge management and the evolution of systems that enable this fusion. To address this fusion, implementation methodologies and tools must address the required capabilities for both, including automation, performance and flexibility for the process side and collaboration, search and retrieval and taxonomy for the knowledge management side (Fig. 11.6).

There is a huge body of knowledge (the blue cloud) that exists in the minds of a company's expert employees. One of the goals of a successful KM programme is to diligently and selectively move the knowledge into the IT infrastructure so that it can be used to improve the execution of key business processes. It should as well enable the development of new process improvements that improve the effectiveness of delivering services (of any type) to your customers, which, of course, is one of the goals of a successful BPM programme.

A key task is the creation of a taxonomy or knowledge map to enable the process. Indexing and search tools can be used to help develop this taxonomy, and collaborations between experts can help in identifying clusters of useful, validated knowledge, as well as other types of content, including documents, business rules, emails and other types of unstructured information. This must be a dynamic process because true knowledge is very perishable and must be constantly refined, recontextualised and validated before being provisioned for use within a business process.

11.16.1   Fusion Architecture

Since both KM and BPM solutions must live within an enterprise service-oriented architecture, it is useful to construct a service-based architectural model for the fused KM/BPM system. The service model presented in the figure below depicts those services that must be provided, to a greater or lesser degree, to include the

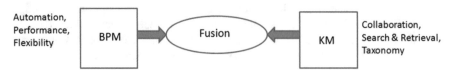

Fig. 11.6 Fusion of process and KM

functionality of BPM, KM and other utility functions that the systems will have to exist alongside. The set of services on the left side of the diagram show these external services. At present, there is no single vendor product that provides all of these required services. Practitioners will find it necessary to architect and build fused KM/BPM systems using conventional techniques for system design and integration (Fig. 11.7).

11.17   Vendor Capabilities

Many BPM, enhanced portal, application server, and workflow products have some of the required components, but seem to be enhancing their tools without a particular strategy in mind. One of the leading BPM vendors, FUEGO, has built adapters for the most ubiquitous document management and portal products, but needs to improve and expand its KM services if it seeks to address this type of need.

There are at least two vendors that are addressing the need for a fused BPM/KM system.

- Kana Resolution is an enhanced call centre package that blends workflow and collaboration/guided search to solve tough service resolution problems.
- Appian is a BPM vendor that has a product vision that appears to reflect the BPM/KM fusion concept but needs to strengthen its process design and simulation capabilities.

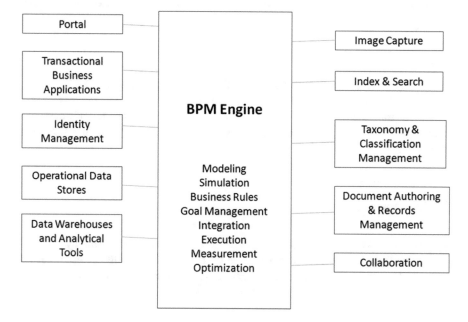

Fig. 11.7 BPM engine

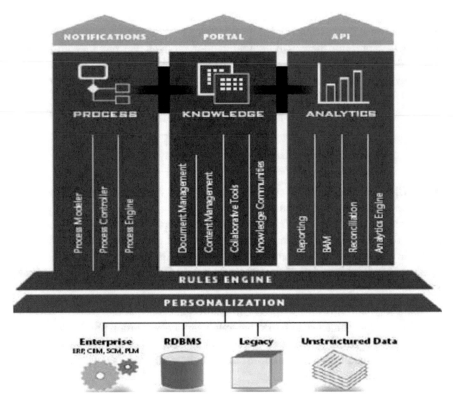

Fig. 11.8 Appian product architecture. *Source*: Introducing the Appian Enterprise 4 BPM Suite, Appian White Paper, 2005

The figure below shows the Appian product architecture (Fig. 11.8).

Appian has integrated its BPM Modeler, process controller and process engine with a separate rules engine and its knowledge components, which include document management, content management, collaborative tools and support of knowledge communities.

11.18   Summary

Automation in knowledge management provides consistency, transparency and predictability features. These features help in pushing the knowledge to the required user by sharing the knowledge at appropriate place. Technology not only helps to share but also helps to publish it without extra overheads. This helps the organisation to create effective metadata for available knowledge and then using search options to make these knowledge available at organisation level. However, a judicious decision needs to be taken regarding the level of investment in technology; this decision needs to be guided by ROI from the investment. This chapter deals with different aspects of usage of technology and the challenges associated with it.

11.19 Appendix

11.19.1 Sample Cost Analysis Template

Cost components		As is		To be	
Categories	Sunk cost	Recurring	Nonrecurring	Recurring	Nonrecurring
Purchase					
• Hardware					
• Software					
• Communications					
• Staff					
• Process					
• Indirect					
Development					
• Hardware					
• Software					
• Communications					
• Staff					
• Process					
• Indirect					
Implementation					
• Hardware					
• Software					
• Communications					
• Staff					
• Process					
• Indirect					
Maintenance					
• Hardware					
• Software					
• Communications					
• Staff					
• Process					
• Indirect					

11.19.2   Sample ROI Calculation Template

	Years					Total
	1	2	3	4	5	
Inflows						
Cost savings						
• Purchase						
• Development						
• Transition						
• Maintenance						
Cost avoidance						
• Purchase						
• Development						
• Transition						
• Maintenance						
Subtotal (1)						A
Outflows						
Costs						
• Purchase						
• Development						
• Transition						
• Maintenance						
Subtotal (2)						B
Return per year (1–2)						
ROI	(A/B)					

11.20   Review Questions

1. What are the factors that affect effective usage of knowledge management?
2. What is the role of knowledge management in present business models?
3. Compare and contrast the type of automation needed for managing knowledge in a public sector and private sector firm within the same domain.
4. Describe the three-tiered KMS architecture.
5. What are the salient features of the US Navy metrics model for calculating the ROI on KM initiatives?
6. What are the prerequisites for any model to calculate the ROI?
7. How can intangibles be quantified? Give some examples.

Appendix

Sample cost analysis template

| Cost components | Sunk cost | As is | | To be | |
Categories		Recurring	Nonrecurring	Recurring	Nonrecurring
Purchase					
• Hardware					
• Software					
• Communications					
• Staff					
• Process					
• Indirect					
Development					
• Hardware					
Software					
• Communications					
• Staff					
• Process					
• Indirect					
Implementation					
• Hardware					
• Software					
• Communications					
• Staff					
• Process					
• Indirect					
Maintenance					
• Hardware					
• Software					

© Springer International Publishing Switzerland 2016
S. Mohapatra et al., *Designing Knowledge Management-Enabled Business Strategies*, Management for Professionals, DOI 10.1007/978-3-319-33894-1

• Communications					
• Staff					
• Process					
• Indirect					

Sample ROI calculation template

	Years					
	1	2	3	4	5	Total
Inflows						
Cost savings						
• Purchase						
• Development						
• Transition						
• Maintenance						
Cost avoidance						
• Purchase						
• Development						
• Transition						
• Maintenance						
Subtotal (1)						A
Outflows						
Costs						
• Purchase						
• Development						
• Transition						
• Maintenance						
Subtotal (2)						B
Return per year (1–2)						
ROI	(A/B)					

Review Questions

1. What are the factors that affect effective usage of knowledge management?
2. What is the role of knowledge management in present business models?
3. Compare and contrast the type of automation needed for managing knowledge in a public sector and private sector firm within the same domain.
4. Describe the three-tiered KMS architecture.
5. What are the salient features of the US Navy metrics model for calculating the ROI on KM initiatives?
6. What are the prerequisites for any model to calculate the ROI?
7. How can intangibles be quantified? Give some examples.

Additional Case Studies

Buckman and KM: Two Sides of the Same Coin!

Buckman Laboratories, a leading speciality chemical company, had pioneered knowledge management under the leadership of then CEO Robert Buckman during the 1990s. The organisation has ever since continually adapted itself to the evolving KM systems and underlying philosophy of knowledge management. The company continues to use knowledge management, a path to increased profitability by addressing knowledge needs of its employee and needs of its customers.

Buckman is known for high-quality innovative chemical supplies and services which are competitively priced. The competitive advantage of Buckman is not only its customer-focused approach and usage of its customer information by sales team across the world but also scientific information related to product processes and development.

Buckman has evolved to sell problem-solving skills and know-how along with its range of chemical products. The company's knowledge-storing and disseminating capabilities about chemical processes which are very complex in nature are important commodities in the market. All of this has been possible by establishing excellent knowledge management practices which have helped in the organisation's continued expansion and development.

Robert Buckman is often known called as the father of modern KM. The company experienced the most well-documented history during the 1990s under his leadership. Connecting employees to help them interact and share thoughts/ideas had been an important concern for the company since long. A number of initiatives had been taken in the past such as:

- 1960s—Distribution of a notebook called "Idea Trap" to allow employees to jot down creative ideas
- 1984—An attempt to create an email system linking all employees across different geographies
- 1986—Laptops for employees based out of remote locations

© Springer International Publishing Switzerland 2016
S. Mohapatra et al., *Designing Knowledge Management-Enabled Business Strategies*, Management for Professionals, DOI 10.1007/978-3-319-33894-1

- 1989—Setting up of Knowledge Transfer Task Force which monitored and sponsored sharing of knowledge within the organisation
- 1992—Implementation of corporate knowledge sharing system called K'Netix

During the early 1990s, Buckman has realised that a product-focused approach alone will not help them sustain in the long run of increasing competition. It was important for the company to become customer focused and serve its customers the right mix of products and processes. This new direction was however dependent on highly specialised knowledge about chemicals, and even more important was spreading this knowledge across the organisation so that all sales teams were completely thorough with this knowledge. Such knowledge would help sales team best cater to the needs of the customers. An efficient knowledge system called K'Netix was established for accurate knowledge of different product combinations for a wide range of customers. K'Netix allowed employees to interact with each other on forums, ask questions and give answers to other people's queries. K'Netix was established on the underlying Buckman Code of Ethics to help in creating an open and trusting knowledge sharing environment for all.

Just 2 years after K'Netix was put in place, technology-enabled courses were made available to train employees. In the next few years, the company integrated its technology capabilities with the capabilities of its workforce to provide a flexible and comprehensive training system with the Bulab Learning Centre. The learning centre was helpful in the following ways:

- Fully informed and trained employees will have a greater potential to innovate
- Fully trained employees will be able to better understand customer needs and hence result in better customer satisfaction
- Creation of a knowledge sharing culture in the organisation
- Greater employee retention rates
- Increasing organisation's capability to adapt to changing environment

Implementation of such knowledge management initiatives in the organisation has led to 50 % rise in sales, 51 % rise in sales per associate and 91 % rise in operating profit per associate.

ShareNet: Knowledge Management Solution for Siemens Information and Communication Networks (ICN) Division

Siemens' ICN division is a global player in providing telecommunications solutions to its customers in more than 100 countries. Traditionally, business was straightforward and simple. Siemens enjoyed a monopoly by maintaining a close relationship with the national telecom which used to be regulated and sold integrated products around the world. But later, during the mid-1990s, the sector underwent a massive transformation as the market became deregulated. Not only did new types of players emerged in the market but also innovation in technology was also happening at a rapid pace with introduction of IP networks in the market. Customers were not just interested in buying boxes but needed complete solutions which included system integration, financing, consulting and other services. There was a significant shift from just product business to a business which has solutions approach and is service focused which led to increased knowledge intensity and complexity of Siemens' business.

The company soon realised that it had to rely heavily on the front lines who are more aware and knowledgeable about the latest market developments. The sales employees had to also act as consultants for the customers. Skills such as network planning, outsourcing, business development, business analysis and so on were in high demand around the world. The biggest challenge for Siemens was to identify the best practices and share them globally. A global knowledge sharing network, ShareNet, was developed to cater to the company's needs.

Today, ShareNet is a close-knit community of more than 18,000 employees of Siemens ICN in departments such as marketing, sales, research and development and business development in more than 80 countries. ShareNet as a platform is used by these experts around the globe to create better customer solutions by sharing and developing their knowledge. Its goal is to identify innovations done at the local level and scale them up to the global level. The platform covers both tacit and explicit knowledge of end-to-end sales value creation process which includes functional and technical solution components, business environment and project know-how. ShareNet helps in developing experience-based knowledge by allowing employees to share personal statements, field experiences, comments and pros and cons for a suggested solution. It also has forums for discussions, community news, chat rooms, etc.

In many geographies, it has been made mandatory for employees to fill a Web-based project questionnaire and, in cases of important project, a milestone. This leads to fulfilment of the fundamental paradigm of this knowledge sharing platform that every user/reader of ShareNet is also a publisher. It allows readers to validate the source of every contribution on the platform, since every contribution is personalised. Employees use ShareNet throughout their sales process for specific solutions, innovative pricing or financing schemes, handling competitors and a lot more.

Siemens defined new functions and roles in the organisation to foster knowledge management efforts as the following:

- Setting up of Knowledge Management Office (KMO) which has a Corporate Knowledge Officer (CKO) and a Corporate Knowledge Management Program Manager.
- Setting up of Knowledge Management Board (KMB) where executives are responsible to define the knowledge management direction, coordinating knowledge management network and developing methodology and processes of knowledge management across the organisation.
- Setting up of Knowledge Management Council (KMC) which includes chief knowledge officers for various groups and regions.

Siemens has invested more than $7.8 million setting up ShareNet which has led to several increased sales worth more than $122 million. But the big question is "Why should employees use ShareNet?"

- ShareNet members save time by getting quick answers to their urgent problems and also make them feel responsible to give back to the community by helping others.
- Often in big organisations, identification of crucial subject matter experts is lost on organisation charts. ShareNet gives them an opportunity to get recognised across the organisation.
- An online incentive system has been put in place where platform members receive points for their valuable contribution which can be redeemed against air miles, free participation in conferences and seminars, etc.

ShareNet helps in skipping the process of reinventing the wheel, and this time is now well spent in building effective and long-lasting relationships with the customer leading to creation of new opportunities.

Further Reading

Alavi M, Leidner DE. Review: knowledge management and knowledge management systems: conceptual foundations and research issues. MIS Quarter. 2001;25(1):107–36.

Allee V. Twelve principles of knowledge management. Train Dev. 1997;51(11):71–4.

Dalkir K. Knowledge management in theory and practice. 2nd ed. Cambridge: The MIT Press; 2011.

Davenport T, Klahr P. Managing customer support knowledge. Calif Manage Rev. 1998;40(3):195–208.

Gupta B, Iyer LS, Aronson JE. Knowledge management: practices and challenges. Ind Manag Data Syst. 2000;100(1):17–21.

Jawadekar WS. Knowledge management: text and cases. New Delhi: Tata Mcgraw Hill Education; 2010.

Holm, J. (2001), Capturing the spirit of knowledge management, paper presented at the American Conference on Information Systems, Boston, MA, August 3–5.

Nonaka and Takeuchi knowledge management cycle. http://hubpages.com/education/Nonaka-and-Takeuchi-knowledge-management-cycle (27 Mar, 2011). Accessed 20 Feb 2014.

Knowledge Capture. http://it.toolbox.com/wiki/index.php/Knowledge_capture (27 Aug, 2008). Accessed 15 Mar 2014.

Knowledge Capture and Knowledge Management. http://www.youtube.com/watch?v=Xc41pMzsUzs (12 Apr, 2010). Accessed 22 Jan 2014.

Wenger E. Supporting communities of practice: a survey of community-oriented technologies. https://guard.canberra.edu.au/opus/copyright_register/repository/53/153/01_03_CP_technology_survey_v3.pdf (Mar, 2001). Accessed 23 Jan 2014.

Brown JS, Duguid P. Organizational learning and communities-of-practice: toward a unified view of working, learning, and innovation. JSTOR; 1991.

Wenger E, Snyder W. Communities of practice: the organizational frontier. Harvard Business Review; 2000.

Lesser E, Everest K. Using communities of practice to manage intellectual capital. Ivey Business Journal. 2001.

Biscozzo M, Corallo A, Elia, G. Building bottom-up ontologies for communities of practice in high-tech firms. Conference: Knowledge-Based Intelligent Information and Engineering Systems, 9th International Conference, KES 2005, Melbourne, Australia, September 14–16, 2005, Proceedings, Part I. doi: 10.1007/11552413_21.

Gannon-Leary P, Fontainha E. Communities of practice and virtual learning communities: benefits, barriers and success factors. eLearning Papers. 2007;5.

© Springer International Publishing Switzerland 2016   193
S. Mohapatra et al., *Designing Knowledge Management-Enabled Business Strategies*, Management for Professionals, DOI 10.1007/978-3-319-33894-1

Wenger E. Communities of practice: learning as a social system. social learning systems and communities of practice. Springer; 2010. pp. 179–98.

http://www.knoco.com/communities-of-practice.htm

Alexander A. Motivation and barriers to participation in virtual knowledge-sharing communities of practice. Journal of Knowledge Management. 1997.

http://www.knoco.com/Knoco%20whitepaper%20_%20selecting%20a%20CoP.pdf

http://www.knoco.com/Knoco%20white%20paper%20-%20evolution%20of%20a%20community.pdf

Levinson M. Knowledge management definition and solutions. 2007. http://www.cio.com/article/40343/knowledge_management_definition_and_solutions. Accessed 24 Jan 2014.

http://kmwiki.wikispaces.com/km+introduction

Chua A, Lam W. Why KM projects fail: a multi-case analysis. J Knowl Manag. 2005;9(3):6–17.

The knowledge-based economy. organisation for economic co-operation and development. Paris; 1996.

Warier S. Knowledge management. Vikas Publishing.

Awad EM, Ghaziri HM. Knowledge management. Pearson; 2007.

Friga PN. Codification strategies in knowledge management processes—learning from simulation. 2000.

Styhre A. Knowledge management beyond codification: knowing as practice/concept. J Knowl Manag. 2003;7:32–40.

Singh H, Zollo M. The impact of knowledge codification, experience trajectories and integration strategies on the performance of corporate acquisitions. Wharton: University of Pennsylvania; 1998.

Liebowitz J. Developing metrics for determining knowledge management success: a fuzzy logic approach. Johns Hopkins University. 2005.

Firestone J. Knowledge management metrics development: a technical approach. 1998. http://www.dkms.com/papers/kmmeasurement.pdf. Accessed 24 Feb 2014.

Kankanhalli A, Tan BCY. Knowledge management metrics: a review and directions for future research. National University of Singapore.

Perez-Soltero A, Barcelo-Valenzuela M, Sanchez-Schmitz G, Martin-Rubio F, Palma-Mendez JT. Knowledge audit methodology with emphasis on core processes. European and Mediterranean Conference on Information Systems (EMCIS) 2006, July 6–7 2006. Costa Blanca, Alicante, Spain.

Sanghani P. Knowledge management implementation: holistic framework based on indian study. Pacific Asia Conference on Information Systems. 2009.

Knowledge management strategy. IFAD. 2007. https://www.ifad.org/documents/10180/ad197dcd-93f9-4e50-ab3dd773619a89e5.

Robertson J.: Developing a Knowledge Management Strategy. KM Column. 2004. http://www.steptwo.com.au/files/kmc_kmstrategy.pdf.

KM consulting methodology overview. http://www.knowledge-management-online.com/KM-consulting-method-overview.html

Lee H. Knowledge management & the role of libraries. http://www.white-clouds.com/iclc/cliej/cl19lee.htm.

Mathew V, Kavitha M. Implementing knowledge management knowledge mapping, matrix and supports. J Knowl Manag. 2009;10(1). http://www.tlainc.com/articl179.htm.

Hurley TA, Green CW. Knowledge management and the non-profit industry: a within and between approach. Journal of Knowledge Management Practice 2005. http://www.tlainc.com/articl79.htm

Are social networking sites knowledge management. 2008. http://kmspace.blogspot.com/2008/04/are-social-networking-sites-knowledge.html.

Curb the recession: welcome to the fast lane of internal social networking-Leverage Software. Leverage Software.

Vivek Paul turns entrepreneur with KineticGlue. http://economictimes.indiatimes.com/news/news-by-company/corporate-announcement/Vivek-Paul-turns-entrepreneur-with-KineticGlue/articleshow/6235092.cms. Economic Times (30 Jul, 2010). Accessed 20 Apr 2014.

O'Reilly T. What is web 2.0. http://oreilly.com/web2/archive/what-is-web-20.html (30 Sep, 2005). Accessed 23 Mar 2014.

Web 2.0 technology primer, blue coat. https://www.bluecoat.com/sites/default/files/documents/files/bcs_tp_Web20_v3b.pdf. Accessed 23 Apr 2014.

Uzzi B, Dunlap, S. How to build your network. Harvard Business Review (Dec, 2005).

Ranganath A. Is web 2.0 aiding in knowledge management—an indian perspective. http://www.scribd.com/doc/24475667/Is-Web-2-0-Aiding-in-Knowledge-Management-An-Indian-Perspective# (Oct, 2009).

Levy M. Web 2.0 implications on knowledge management. J Knowl Manag. 2009;13(1):120–34.

Zhang L, Tu W. Six degrees of separation in online society. http://journal.webscience.org/147/2/websci09_submission_49.pdf (Mar, 2009).

Panckhurst R, Marsha D. Communities of practice. Using the open web as a collaborative learning platform. iLearning Forum. 2008. https://hal.archives-ouvertes.fr/hal-00291874/document.

Leone S, Grossniklaus M, Norrie MC. Architecture for integrating desktop and web 2.0 data management. https://globis.ethz.ch/?pubdownload=543.

Yasin R. Knowledge management in the cloud: catalyst for open government? http://fcw.com/articles/2010/05/03/knowledge-management-cloud-computing.aspx (3 May, 2010). Accessed 14 Mar 2014.

Fitzgerald M. Why social computing aids knowledge management. http://www.cio.com/article/395113/Why_Social_Computing_Aids_Knowledge_Management_?page=3&taxonomyId=3000 (13 Jun, 2008). Accessed 12 Mar 2014.

Stryer P. Understanding data centres and cloud computing. http://viewer.media.bitpipe.com/1078177630_947/1267474882_422/WP_DC_DataCenterCloudComputing1.pdf. Accessed 15 Mar 2014.

Knowledge management—emerging perspectives. http://www.systems-thinking.org/kmgmt/kmgmt.htm. Accessed 15 Mar 2014.

7 things you should know about cloud computing. http://net.educause.edu/ir/library/pdf/EST0902.pdf (Aug, 2009).

McEvoy N. Cloud computing roadmap. http://cloudventures.files.wordpress.com/2010/11/cloud_computing_roadmap.pdf. Accessed 5 Mar 2014.

Chua A. Knowledge management system architecture: a bridge between KM consultants and technologists. Int J Inf Manag. 2004;24(1):87–98.

Kucza T. Knowledge management process model. Technical Research Centre of Finland; 2001. http://www.vtt.fi/inf/pdf/publications/2001/P455.pdf.

Delivering the business benefits of service-oriented architecture. CSC White Paper. 2005. http://assetsdev1.csc.com/au/downloads/10510_1.pdf.

Introducing the appian enterprise 4 BPM suite. Appian White Paper. 2005.

The fusion of process and knowledge management. BP Trends. http://www.bptrends.com/publicationfiles/09-05%20WP%20Fusion%20Process%20KM%20-%20Records.pdf (Sep, 2005).

Service resolution management. KANA white paper. 2005.

Bonett M. Personalization of web services: opportunities and challenges. www.ariadne.ac.uk/issue28/personalization/. Accessed 15 Apr 2014.

Estimating return on investment for knowledge management initiatives: an information technology perspective. BEI Consulting. https://www.k4health.org/sites/default/files/EstimatingROI.pdf.

Tobin T. The insider's guide to knowledge management ROI—quantifying knowledge-enabled customer service and support. http://www.rcc.gov.pt/sitecollectiondocuments/whitepaper-roi-2004.pdf. ServiceWare White Paper (Feb, 2004).

Definition of knowledge management. http://www.knowledge-management-online.com/Definition-of-Knowledge-Management.html. Accessed 7 Jan 2014.

Levinson M. Knowledge management definition and solutions. http://www.cio.com/article/40343/
    Knowledge_Management_Definition_and_Solutions. Accessed 7 Jan 2014.
http://www.kmbestpractices.com/buckman-laboratories.html
KM project successful case study of Buckman Laboratories. http://www.ravi.kahlon.co/2013/01/
    km-project-successful-case-study-of.html. Accessed 15 Jan 2014.
Ellis MS, Rumizen R. The evolution of KM at Buckman Laboratories. http://www.providersedge.
    com/docs/km_articles/The_Evolution_of_KM_at_Buchman_Labs.pdf. Accessed 20 Jan 2014.

Printed in the United States
By Bookmasters